Macmillan/McGraw-Hill • Glencoe

2

Math Triumphs

Book 1: Number and Operations

Authors

Basich Whitney • Brown • Dawson • Gonsalves • Silbey • Vielhaber

**Macmillan/McGraw-Hill
Glencoe**

Photo Credits

All coins photographed by United States Mint.
All bills photographed by Michael Houghton/StudiOhio.
Cover Fancy Photography/Veer; **vi–viii** Mazer Creative Services; **1** The McGraw-Hill Companies; **3** Mazer Creative Services; **6** Getty Images, (b)Mazer Creative Services; **7** Getty Images; **12** (1)Jupiterimages, (2)Siede Preis/Getty Images, (3, 4, 5)Getty Images; **13, 17** Mazer Creative Services; **23** Eclipse Studios; **30** Siede Preis/Getty Images; **32** (t)Alamy, (b)Getty Images; **37, 41** Mazer Creative Services; **44** G.K. & Vikki Hart/Getty Images; **46** (l)RubberBall, (r)David Katzenstein/CORBIS; **49** David Young-Wolff/Getty Images; **51** Getty Images; **53** Jules Frazier/Getty Images; **59** Siede Preis/Getty Images; **61** Mazer Creative Services; **62, 64, 67** Ken Cavanagh/The McGraw-Hill Companies; **68** Getty Images; **70** (1)PunchStock, (2, 4–6)G.K. & Vikki Hart/Getty Images, (3)Getty Images; **71** Barbara Penoyar/Getty Images; **77** Getty Images; **78** (t)Getty Images, (b)PunchStock; **82** (t)G.K. & Vikki Hart/Getty Images, (b)Siede Preis/Getty Images; **83** Jules Frazier/Getty Images; **84** Getty Images; **85** Richard Hutchings; **86** (t)Michael Newman/PhotoEdit, (b)Dex Image/CORBIS; **94** (l)Eclipse Studios, (r)Alamy; **97** Mazer Creative Services; **99** Getty Images; **102** (l)Getty Images, (r)G.K. & Vikki Hart/Getty Images; **103** (l)Getty Images, (r)Mazer Creative Services; **105** Ken Cavanagh/The McGraw-Hill Companies; **111** Alamy; **117** Chris Howes/Alamy; **119, 123** Michael Newman/PhotoEdit; **126** PunchStock; **132** (tl)Eclipse Studios, (tr)Alamy, (b)Bobbi Tull/Getty Images.

The McGraw-Hill Companies

 Macmillan/McGraw-Hill
Glencoe

Send all inquiries to:
Macmillan/McGraw-Hill • Glencoe/McGraw-Hill
8787 Orion Place
Columbus, OH 43240-4027

ISBN: 978-0-07-888195-4
MHID: 0-07-888195-1

Printed in the United States of America.

4 5 6 7 8 9 10 066 16 15 14 13 12 11 10 09

Math Triumphs
Grade 2, Book 1

Math Triumphs

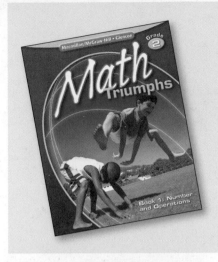
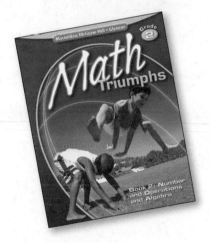
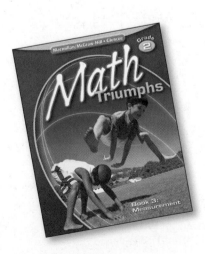

Authors and Consultants

CONSULTING AUTHORS

Frances Basich Whitney
Project Director, Mathematics K–12
Santa Cruz County Office of Education
Capitola, California

Kathleen M. Brown
Math Curriculum Staff Developer
Washington Middle School
Long Beach, California

Dixie Dawson
Math Curriculum Leader
Long Beach Unified
Long Beach, California

Philip Gonsalves
Mathematics Coordinator
Alameda County Office of Education
Hayward, California

Robyn Silbey
Math Specialist
Montgomery County Public Schools
Gaithersburg, Maryland

Kathy Vielhaber
Mathematics Consultant
St. Louis, Missouri

CONTRIBUTING AUTHORS

Viken Hovsepian
Professor of Mathematics
Rio Hondo College
Whittier, California

FOLDABLES Study Organizer **Dinah Zike**
Educational Consultant
Dinah-Might Activities, Inc.
San Antonio, Texas

CONSULTANTS

Assessment

Donna M. Kopenski, Ed.D.
Math Coordinator K–5
City Heights Educational Collaborative
San Diego, California

Instructional Planning and Support

Beatrice Luchin
Mathematics Consultant
League City, Texas

ELL Support and Vocabulary

ReLeah Cossett Lent
Author/Educational Consultant
Alford, Florida

Reviewers

Each person below reviewed at least two chapters of the Student Study Guide, providing feedback and suggestions for improving the effectiveness of the mathematics instruction.

Dana M. Addis
Teacher Leader
Dearborn Public Schools
Dearborn, MI

Renee M. Blanchard
Elementary Math Facilitator
Erie School District
Erie, PA

Jeanette Collins Cantrell
5th and 6th Grade Math Teacher
W.R. Castle Memorial Elementary
Wittensville, KY

Helen L. Cheek
K–5 Math Specialist
Durham Public Schools
Durham, NC

Mercy Cosper
1st Grade Teacher
Pershing Park Elementary
Killeen, TX

Bonnie H. Ennis
Math Coordinator
Wicomico County Public Schools
Salisbury, MD

Sheila A. Evans
Instructional Support Teacher–Math
Glenmount Elementary/Middle School
Baltimore, MD

Lisa B. Golub
Curriculum Resource Teacher
Millennia Elementary
Orlando, FL

Donna Hagan
Program Specialist–Special Programs
    Department
Weatherford ISD
Weatherford, TX

Russell Hinson
Teacher
Belleview Elementary
Rock Hill, SC

Tania Shepherd Holbrook
Teacher
Central Elementary School
Paintsville, KY

Stephanie J. Howard
3rd Grade Teacher
Preston Smith Elementary
Lubbock, TX

Rhonda T. Inskeep
Math Support Teacher
Stevens Forest Elementary School
Columbia, MD

Albert Gregory Knights
Teacher/4th Grade/Math Lead Teacher
Cornelius Elementary
Houston, TX

Barbara Langley
Math/Science Coach
Poinciana Elementary School
Kissimmee, FL

David Ennis McBroom
Math/Science Facilitator
John Motley Morehead Elementary
Charlotte, NC

Jan Mercer, MA; NBCT
K–5 Math Lab Facilitator
Meadow Woods Elementary
Orlando, FL

Rosalind R. Mohamed
Instructional Support Teacher–Math
Furley Elementary School
Baltimore, MD

Patricia Penafiel
Teacher
Phyllis Miller Elementary
Miami, FL

Lindsey R. Petlak
2nd Grade Instructor
Prairieview Elementary School
Hainesville, IL

Lana A. Prichard
District Math Resource Teacher K–8
Lawrence Co. School District
Louisa, KY

Stacy L. Riggle
3rd Grade Spanish Magnet Teacher
Phillips Elementary
Pittsburgh, PA

Wendy Scheleur
5th Grade Teacher
Piney Orchard Elementary
Odenton, MD

Stacey L. Shapiro
Teacher
Zilker Elementary
Austin, TX

Kim Wilkerson Smith
4th Grade Teacher
Casey Elementary School
Austin, TX

Wyolonda M. Smith, NBCT
4th Grade Teacher
Pilot Elementary School
Greensboro, NC

Kristen M. Stone
3rd Grade Teacher
Tanglewood Elementary
Lumberton, NC

Jamie M. Williams
Math Specialist
New York Mills Union Free School District
New York Mills, NY

Contents

CHAPTER 1 Whole Numbers

CHAPTER 2 — Place Value

Contents

CHAPTER 3 Compare and Order Whole Numbers

Home Connection

English

Dear Family,
Today our class started **Chapter 1, Whole Numbers.** In this chapter, I will learn to read, write, and recognize numbers up to 500. I will also learn to skip count by 2s, 5s, and 10s.

Love, _____

Spanish

Estimada familia,
Hoy en clase comenzamos **el Capítulo 1, titulado Números Enteros.** En este capítulo aprenderé a leer, escribir, y reconocer los números del 1 al 500. También aprenderé a contar de 2 en 2, de 5 en 5 y de 10 en 10.

Cariños, _____

Help at Home

You can work with your child to count the numbers both forward and backward. Show them numbers at home, such as on a calendar, in magazines, or in advertisements. Ask your child to tell you what numbers he or she sees.

Ayude en casa

Usted puede trabajar con su hijo(a) para contar hacia adelante y hacia atrás. Muéstrele los números en casa, como en el calendario, en revistas o en anuncios. Pídale a su hijo(a) que le diga qué números ve a su alrededor.

Math Online ▷ Take the chapter Get Ready quiz at macmillanmh.com.

Name _____

Get Ready

Write the missing numbers.

1 1, _____, _____, 4, 5

2 7, 8, _____, _____, _____

3 Circle the number that comes before 6.

5   9

4 Circle the number that comes after 9.

6   10

5 Count. Write each number.

_____   _____   _____   _____

6 Circle the boy in shirt number seven.

7 Write the value of each model.

_____   _____   _____

8 Angela counts the fish in her tank.

How many fish are there? _____

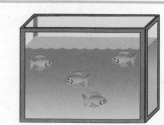

Numbers 0 to 20

Key Concept

How many cubes are shown?

Count to tell how many.

Count: 1   2   3   4   5   6   7   8   9   10   11

There are 11 cubes.

There are eleven cubes.

Vocabulary

count

1   2   3
one   two   three

number   tells how many
1, 2, 3, 4, 5, 6, 7, 8, 9, 10, . . .

You can count to tell how many.

You can draw pictures to show a **number**.

GO on

Example

Count the flowers. Write the number and the number name.

Step 1   Start at 1. Count the flowers.

Step 2   Write the numbers as you count.
1, 2, 3, 4, 5, 6, 7, 8, 9, 10, 11, 12, 13, 14, 15, 16

Step 3   The ending number tells how many.
The ending number is 16.

Answer   There are 16, or sixteen, flowers.

Step-by-Step Practice

Count the leaves. Write the number and the number name.

Step 1   Start at ___1___. Count the leaves.

Step 2   Write the numbers as you count.

___1___, ___2___, ___3___, _____, _____,

_____, _____, _____, _____

Step 3   The ending number tells how many.

The ending number is _____.

Answer   There are _____, or _____, leaves.

Name _____

 Guided Practice

Count. Write the number and the number name.

1 [books images]

_____ books or _____ books

2 [pencils images]

_____ pencils or _____ pencils

3 Draw ☐ to show the number 10.

Problem-Solving Practice

4 Sam has fourteen baseball cards.
   Show fourteen baseball cards.

   Understand   Underline key words.

   Plan   Draw a picture.

   Solve   Draw 14 baseball cards.

   Check   Count the baseball cards you drew.

GO on

Count. Write the number and the number name.

5

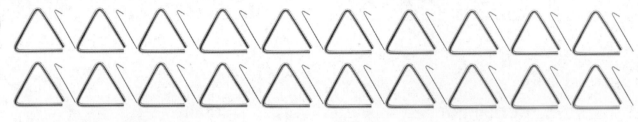

_____ recorders or _____ recorders

6

_____ triangles or _____ triangles

---

7 Draw ♡ to show the number thirteen.

---

8 **WRITING IN ►MATH** Look at the picture. Isabel can choose any instrument. How many choices does she have? Explain.

_____

_____

Vocabulary Check Complete.

9 A _____ tells how many.

Name _____

Numbers 0 to 50

Key Concept

Count the apples to find how many.

twenty-one,   twenty-two,   twenty-three,   twenty-four

Count the groups of ten apples.
There are twenty apples in groups of ten.

How many apples are left? four apples

There are 24 apples. There are twenty-four apples.

Vocabulary

count

1
one

2
two

3
three

2 groups of ten equal twenty.

3 groups of ten equal thirty.

4 groups of ten equal forty.

5 groups of ten equal fifty.

1	2	3	4	5	6	7	8	9	10
11	12	13	14	15	16	17	18	19	20
21	22	23	24	25	26	27	28	29	30
31	32	33	34	35	36	37	38	39	40
41	42	43	44	45	46	47	48	49	50

GO on

Example

Count the crayons. Write the number and the number name.

Step 1   Count the groups of ten crayons. 10, 20, 30
There are 30 crayons in groups of ten.

Step 2   Count the crayons left. 1, 2, 3, 4, 5, 6, 7
There are 7 crayons left.

Answer   There are 37 crayons.
There are thirty-seven crayons.

Step-by-Step Practice

Count the dimes. Write the number and the number name.

Step 1   Count the groups of ten dimes.

10 , _____, _____, _____

There are _____ dimes in groups of ten.

Step 2   Count the dimes left. _____, _____, _____

There are _____ dimes left.

Answer   There are _____ dimes.

There are _____ dimes.

Name _____

▶ Guided Practice

Count. Write the number and the number name.

1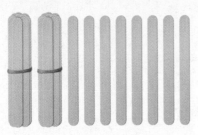

_____ craft sticks or

_____ craft sticks

2

_____ crayons or

_____ crayons

Problem-Solving Practice

3 Miss Lincoln's second-grade class had a pizza party. She ordered 5 pizzas. How many slices are there in all?

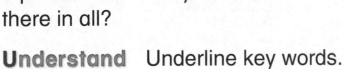

Understand   Underline key words.

Plan   Count the groups of ten.

Solve   There are _____ pizzas.

_____ groups of ten equal

_____ or _____.

There are _____ or _____ slices in all.

Check   Count each of the slices.

GO on

▶ Practice on Your Own

Write each number.

4 thirty-eight

_____

5 forty-six

_____

6 twenty-two

_____

Write each number name.

7 25

_____

8 49

_____

9 31

_____

10 Count. Write the number and the number name.

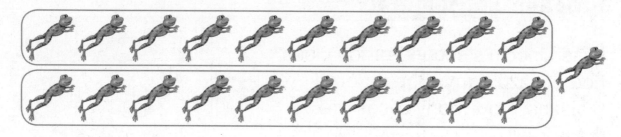

_____ frogs or _____ frogs

11 WRITING IN ▶MATH   Clifford wants to write the number name for 35. He writes twenty-five. Is Clifford correct? Explain.

_____

_____

Vocabulary Check Complete.

12 I can _____ to tell a number.

STOP

Name _____

Progress Check 1 (Lessons 1-1 and 1-2)

Write each number.

1 seven

2 eighteen

3 thirty-nine

_____

_____

_____

Write each number name.

4 12

5 26

6 45

_____

_____

_____

7 Count. Write the number and the number name.

_____ craft sticks or _____ craft sticks

8 Mrs. Brody has fourteen students in her class.

Draw 😊 to show the number fourteen.

9 Kareem writes the number names for 34, 35, and 36.
What number name will Kareem write next? Explain.

thirty-four, thirty-five, _____

_____

GO on

Name _____

I Spy a Butterfly

Diego saw a butterfly flying outside his window.

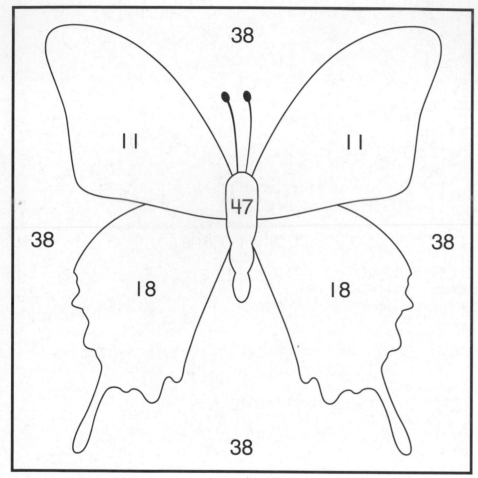

Write each number. Color the picture.

eleven _____   eighteen _____

Color eleven ⬤ .   Color eighteen ⬤ .

forty-seven _____   thirty-eight _____

Color forty-seven ⬤ . Color thirty-eight ⬤ .

Name _____

Numbers 0 to 100

Key Concept

Hundred Chart

1	2	3	4	5	6	7	8	9	10
11	12	13	14	15	16	17	18	19	20
21	22	23	24	25	26	27	28	29	30
31	32	33	34	35	36	37	38	39	40
41	42	43	44	45	46	47	48	49	50
51	52	53	54	55	56	57	58	59	60
61	62	63	64	65	66	67	68	69	70
71	72	73	74	75	76	77	78	79	80
81	82	83	84	85	86	87	88	89	90
91	92	93	94	95	96	97	98	99	100

50 or fifty
60 or sixty
70 or seventy
80 or eighty
90 or ninety
100 or one hundred

You can use a **hundred chart** to read and count numbers to one hundred, or 100.

Vocabulary

hundred chart   a chart that shows numbers 1 to 100

61	62	**63**	64	65	66	67	68	69	70

Ayita wrote the missing number.
The number 63 is between 62 and 64.

GO on

Write the missing numbers.

81	82	83		85	86	87		89	90

Step 1   Count from 81 to 90.
81, 82, 83, 84, 85, 86, 87, 88, 89, 90

Step 2   The number 84 is between 83 and 85.

Step 3   The number 88 is between 87 and 89.

Answer   Write 84 and 88 in the correct boxes.

81	82	83	84	85	86	87	88	89	90

Step-by-Step Practice

Write the missing numbers.

51	52	53	54	55	56		58	59	

Step 1   Count from 51 to 60.

__51__ , __52__ , _____ , _____ , _____ ,

_____ , _____ , _____ , _____ , _____

Step 2   The number _____ is between 56 and 58.

Step 3   The number _____ is after 59.

Answer   Write _____ and _____ in the correct boxes.

51	52	53	54	55	56		58	59	

Name _____

▶ Guided Practice

Write the missing numbers.

1

71		73	74	75		77	78	79	80

2

	92	93	94	95		97	98	99	100

Problem-Solving Practice

3 Alejandra used chalk to draw a number chart on the sidewalk. The rain washed away some of the numbers. Write the missing numbers.

61	62	63	64		66	67		69	70
71		73		75	76	77	78	79	80

Understand   Underline key words.

Plan   Count.

Solve   Count from _____ to _____.
Write the missing numbers as you count.

Check   Use a hundred chart to check your answer.

GO on

▶ Practice on Your Own

Write the missing numbers.

4

	72		74	75	76	77	78	79	80

5

51	52	53	54			57	58	59	60

6

61	62	63	64	65	66	67	68		

7

91		93	94	95	96		98	99	100

---

8 **WRITING IN ▶MATH** Harold knows that a secret number comes just before 88. Use the chart to help Harold find the secret number.

81	82	83	84	85	86	87	88	89	90

What is the secret number? How do you know?

_____

_____

Vocabulary Check Complete.

9 A _____ is a chart that shows numbers 1 to 100.

STOP

Name _____

Numbers 0 to 200

Key Concept

You can use a **number line** to read and count.

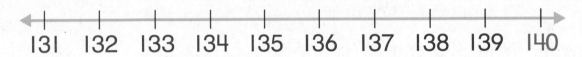

101 102 103 104 105 106 107 108 109 110

The number 107 is between 106 and 108.

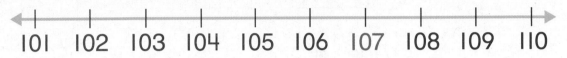

131 132 133 134 135 136 137 138 139 140

The number 140 comes after 139.

Vocabulary

number line   a line with number labels

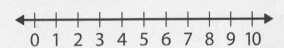

0 1 2 3 4 5 6 7 8 9 10

Ernesto wrote the missing number on the number line. 68 is between 67 and 69. So, 168 is between 167 and 169.

GO on

Example

Write the missing numbers.

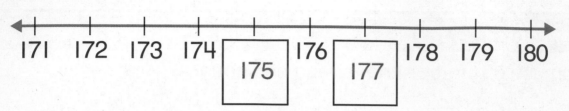

Step 1   Count from 171 to 180.

171, 172, 173, 174, 175, 176, 177, 178, 179, 180

Step 2   The number 175 is between 174 and 176.

Step 3   The number 177 is between 176 and 178.

Answer   Write 175 and 177 in the correct boxes.

Step-by-Step Practice

Write the missing numbers.

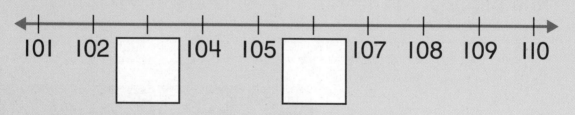

Step 1   Count from 101 to 110.

101, 102, _____, _____, _____,

_____, _____, _____, _____, _____

Step 2   The number _____ is between 102 and 104.

Step 3   The number _____ is is between 105 and 107.

Answer   Write _____ and _____ in the correct boxes.

Name _____

▶ Guided Practice

Write the missing numbers.

1

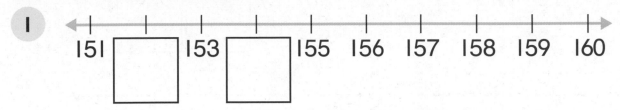

151 | ☐ | 153 | ☐ | 155 | 156 | 157 | 158 | 159 | 160

2

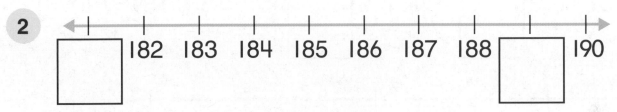

☐ | 182 | 183 | 184 | 185 | 186 | 187 | 188 | ☐ | 190

Problem-Solving Practice

3 Alma is reading a book. The page just after 145 is missing. What page is missing from Alma's book?

Understand   Underline key words.

Plan   Use a number line.

Solve   Circle 145. What number comes next?

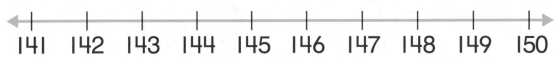

141 142 143 144 145 146 147 148 149 150

The page missing from Alma's book is _____.

Check   Use number patterns.
What number comes after 45?
What number should come after 145?

GO on

▶ Practice on Your Own

Write the missing numbers.

4

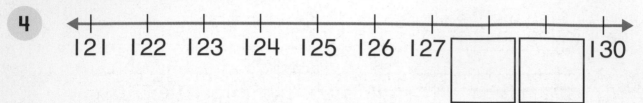

121 122 123 124 125 126 127 ☐ ☐ 130

5

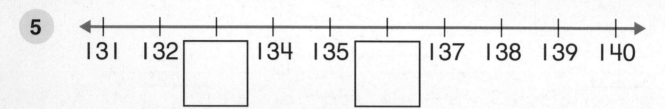

131 132 ☐ 134 135 ☐ 137 138 139 140

6

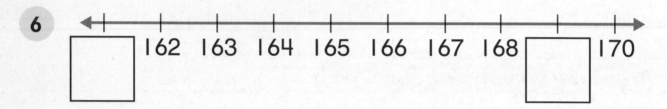

☐ 162 163 164 165 166 167 168 ☐ 170

7

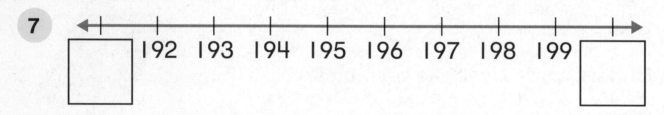

☐ 192 193 194 195 196 197 198 199 ☐

---

8 **WRITING IN ▶MATH** Ricardo counts,
"155, 156, 157, 158, 159, 161, 162, 163, 164, 165."
What number does Ricardo forget? Explain.

_____

_____

Vocabulary Check Complete.

9 A _____ is a line with number labels.

STOP

20 twenty

Progress Check 2 (Lessons 1-3 and 1-4)

Write the missing numbers.

1

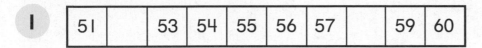

| 51 | | 53 | 54 | 55 | 56 | 57 | | 59 | 60 |

2

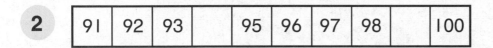

| 91 | 92 | 93 | | 95 | 96 | 97 | 98 | | 100 |

3

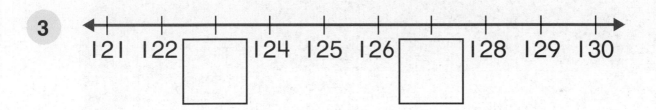

121 122 ☐ 124 125 126 ☐ 128 129 130

4

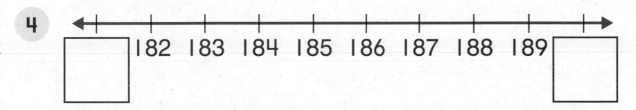

☐ 182 183 184 185 186 187 188 189 ☐

5 Larry collected cans. His goal was to collect 165 cans. He collected 1 more can than his goal. How many cans did Larry collect? Explain.

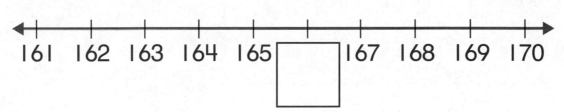

161 162 163 164 165 ☐ 167 168 169 170

«« Replay   Name Game

What is Ralph the dog's last name?
Match each letter with its missing
number to find out.

51	52	53	54	55	56	57	58	59	60
61	62	63	64	65	66	67	68	69	70
71	72	73	74	75	76	77	78	79	80
K	82	83	84	85	86	**E**	88	89	90
91	92	93	94	95	96	97	98	99	100
101	102	**R**	104	105	106	107	108	109	110
111	112	113	114	115	116	117	**R**	119	120
121	122	123	124	125	126	127	128	129	130
131	132	133	**B**	135	136	137	138	139	140
141	142	143	144	145	146	147	148	149	150
151	152	153	154	155	156	157	158	159	160
161	162	163	164	165	166	167	168	169	170
171	172	173	174	175	176	177	178	179	180
181	182	183	184	185	186	187	188	**A**	190
191	192	193	194	195	196	197	198	199	200

134   189   103   81   87   118

Name _____

Numbers 0 to 500

Key Concept

You can use **base-ten blocks** to model numbers through 500.

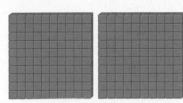

2
two hundred

3
thirty-

7
seven

The blocks show 237.

Vocabulary

base-ten blocks   a set of blocks used to show place value

I know how to model the number 24. So, I know how to model the number 324.

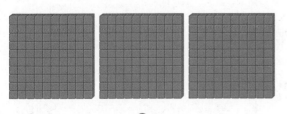

3
three hundred

2
twenty-

4
four

GO on

Example

Count. Write the number and the number name.

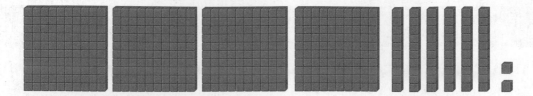

Step 1   Look at the model.

Step 2   There are 4 hundreds flats, or four hundred.

Step 3   There are 6 tens rods, or sixty.

Step 4   There are 2 ones cubes, or two.

Answer   The number is 462, or four hundred sixty-two.

Step-by-Step Practice

Count. Write the number and the number name.

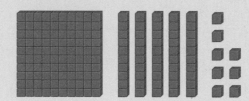

Step 1   Look at the model.

Step 2   There is _____ hundreds flat, or

_____.

Step 3   There are _____ tens rods, or _____.

Step 4   There are _____ ones cubes, or _____.

Answer   The number is _____, or

_____ _____ _____ - _____.

Name _____

▶ Guided Practice

Count. Write the number and the number name.

1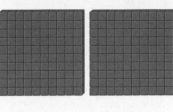

_____2_____   _____   _____

_____   _____   nine

Problem-Solving Practice

2 Abe makes this model. What number does he model?

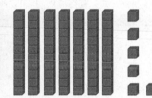

Understand   Underline key words.

Plan   Count the base-ten blocks.

Solve   _____ hundreds, _____ tens,

and _____ ones.

Abe models the number _____.

Check   Use a pattern. Skip count by hundreds and tens. Count on by ones.

GO on

▶ Practice on Your Own

Count. Write the number and the number name.

3

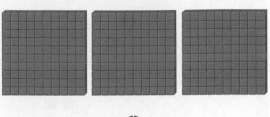 

_____3_____   _____   _____

_____   _____   _____five_____

4

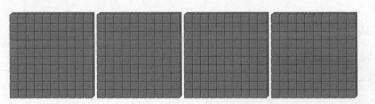 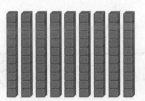

_____   _____   _____

_____ ninety-_____ _____

---

5 **WRITING IN ▶MATH**   The next to last page in a book is 398. How many pages are in the book? Explain.

_____

_____

_____

Vocabulary Check   Complete.

6 _____ are used to show place value.

STOP

26   twenty-six

Name _____

Skip Count by 2s

Key Concept

You can use pennies to **skip count** by 2s.

2   4   6   8   10

You can also use a number line to skip count by 2s.

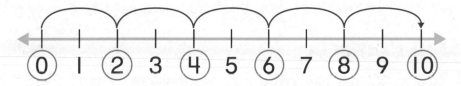

Start at 0. Skip count by 2s. You count 2, 4, 6, 8, and 10.

Vocabulary

skip count to count objects in equal groups of two or more

pattern an order that a set of objects or numbers follows over and over

The numbers 0, 2, 4, 6, 8, and 10 form a **pattern**.
They get larger by 2.

GO on

Example

Skip count by 2s. Write the missing numbers.

10, ____, 14, 16, ____, 20

Step 1   Start at 10 on the number line.
Skip count 2. Land on 12.

Step 2   Start at 12. Skip count. Land on 14.
Skip count. Land on 16.
Skip count. Land on 18.

Answer   The missing numbers are 12 and 18.

Step-by-Step Practice

Skip count by 2s. Write the missing numbers.

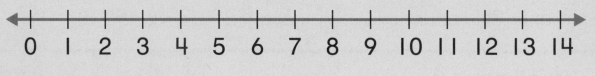

4, 6, _____, 10, _____, 14

Step 1   Start at ___4___ on the number line.

Skip count 2. Land on _____.

Skip count. Land on _____.

Step 2   Start at _____. Skip count. Land on _____.

Skip count. Land on _____.

Answer   The missing numbers are _____ and _____.

Name _____

▶ Guided Practice

Skip count by 2s. Write the missing numbers.

1

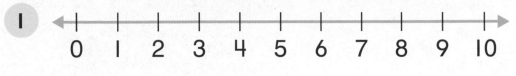

2, _____, 6, 8, _____

2

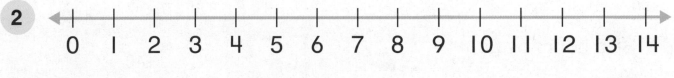

6, 8, _____, _____, 14

Problem-Solving Practice

3 Margie has five pairs of gloves.
How many gloves does Margie have in all?

Understand   Underline key words.

Plan   Skip count by 2s.

Solve   Count 2 for each .

_____, _____, _____, _____, _____

Margie has _____ gloves in all.

Check   Use a number line to skip count by 2s.

GO on

▶ Practice on Your Own

Skip count by 2s. Write the missing numbers.

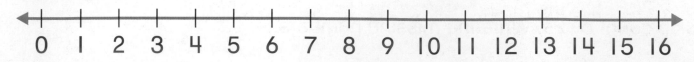

0 1 2 3 4 5 6 7 8 9 10 11 12 13 14 15 16

4 4, _____, 8, _____, _____

5 _____, 10, _____, _____, _____

---

6 Skip count by 2s.

__2__, _____, _____, _____, _____, _____ bunny ears

---

7 ✏️ **WRITING IN ▶MATH** Shani counts the pinecones in her yard. How many pinecones are in Shani's yard? Explain.

_____, _____, _____, _____, _____

_____

_____

Vocabulary Check Complete.

8 10, 12, 14, 16, 18
The numbers get larger by 2.

The numbers form a _____.

30 thirty

STOP

Name _____

Progress Check 3 (Lessons 1-5 and 1-6)

1 Count. Write the number and the number name.

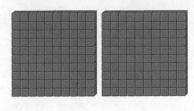 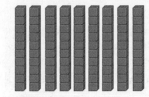

__2__   _____   _____

__two hundred__ _____ _____

---

2 Skip count by 2s. Write the missing numbers.

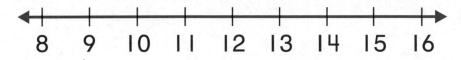

8   9   10   11   12   13   14   15   16

_____, 10, _____, _____, 16

---

3 Skip count by 2s.

_____, _____, _____, _____, _____,

---

4 A grocery store shelf has soup cans. They form a pattern. How many soup cans will be next? Skip count by 2s. Explain.

_____

Name _____

«« **Replay**   **Home Run Mystery**

How many home runs did Tomás hit?
Find each missing number.

1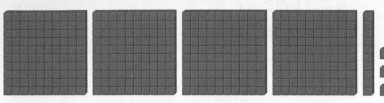

_____

2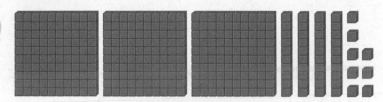

_____

3 4, _____, 8, 10, 12

4 _____, 10, 12, 14, 16

5 10, _____, 14, 16, 18

6 12, 14, 16, _____, 20

7 _____, 2, 4, 6, 8

8 6, _____, 10, 12, 14

9 When you skip count by 2, the numbers always

get larger by _____.

Find each answer in the grid.
Connect the dots in order.

What number is made?

Tomás hit _____ home runs.

344	348	486
413	15	6
10	12	8
18	4	499
0	8	2

Name _____

Skip Count by 5s

Key Concept

1	2	3	4	**5**	6	7	8	9	**10**
11	12	13	14	**15**	16	17	18	19	**20**
21	22	23	24	**25**	26	27	28	29	**30**
31	32	33	34	**35**	36	37	38	39	**40**
41	42	43	44	**45**	46	47	48	49	**50**

Start at 5. Skip count by 5s to 50.

5, 10, 15, 20, 25, 30, 35, 40, 45, 50

The numbers form a pattern. They end in 0 or 5.

10, 15, 20, 25, . . .

Vocabulary

nickel   5¢ or 5 cents

heads   tails

Kanya has 6 **nickels**. Skip count by 5s. How many cents does she have?

_____, _____, _____, _____, _____, _____

Kanya has 30 cents.

GO on

Example

Skip count by 5s. Write the missing numbers.

21	22	23	24	25	26	27	28	29	30
31	32	33	34	35	36	37	38	39	40
41	42	43	44	45	46	47	48	49	50

25, 30, ____, 40, ____, 50

Step 1   Start at 25. Skip count by 5s to 50.
25, 30, 35, 40, 45, 50

Step 2   Write the missing numbers.

Answer   The missing numbers are 35 and 45.

Step-by-Step Practice

Skip count by 5s. Write the missing numbers.

1	2	3	4	5	6	7	8	9	10
11	12	13	14	15	16	17	18	19	20
21	22	23	24	25	26	27	28	29	30

5, 10, _____, 20, _____, 30

Step 1   Start at _____. Skip count by 5s to _____.

_____, _____, _____, _____, _____, _____

Step 2   Write the missing numbers.

Answer   The missing numbers are _____ and _____.

Name _____

 **Guided Practice**

Skip count by 5s. Write the missing numbers.

1

11	12	13	14	15	16	17	18	19	20
21	22	23	24	25	26	27	28	29	30
31	32	33	34	35	36	37	38	39	40

15, _____, 25, _____, 35

2

21	22	23	24	25	26	27	28	29	30
31	32	33	34	35	36	37	38	39	40
41	42	43	44	45	46	47	48	49	50

25, 30, _____, _____, 45

Problem-Solving Practice

3 There are 5 fingers on a hand.
How many fingers are on 6 hands?

Understand Underline key words.

Plan Skip count by 5s.

Solve Count 5 for each hand.

_____, _____, _____, _____, _____, _____ fingers

There are _____ fingers on 5 hands.

Check Use 6 nickels and skip count by 5s.

GO on

▶ Practice on Your Own

Skip count by 5s. Write the missing numbers.

4 5, 10, _____, _____, 25

5 20, 25, _____, 35, _____

1	2	3	4	5	6	7	8	9	10
11	12	13	14	15	16	17	18	19	20
21	22	23	24	25	26	27	28	29	30
31	32	33	34	35	36	37	38	39	40

Skip count by 5s.

6

_5_, _____, _____, _____, _____, _____, _____ balls

7

_____, _____, _____, _____, _____, _____ cookies

8 ✏ **WRITING IN** ▶ **MATH** Tamara has 4 nickels. How many cents does she have? Skip count by 5s. Explain.

_____, _____, _____, _____ cents

_____

Vocabulary Check   Complete.

9 A _____ is equal to 5 cents.

STOP

Name _____

Skip Count by 10s

Key Concept

When you skip count by 10s, you count the numbers
in the chart.

Number	Number Name		Number	Number Name
10	ten		60	sixty
20	twenty		70	seventy
30	thirty		80	eighty
40	forty		90	ninety
50	fifty		100	one hundred

The numbers form a pattern. They all end in 0.
10, 20, 30, 40, 50, 60, 70, 80, 90, 100

Vocabulary

dime   10¢ or 10 cents.

You can use **dimes** to skip count by 10s.

10¢   20¢   30¢

40¢   50¢   60¢

GO on

Example

Skip count by 10s. Write the missing numbers.

60, _____, 80, _____, 100

sixty,   seventy,   eighty,   ninety,   one hundred

Step 1   Use the chart. Start at 60. Skip count by 10s.
The number 70 comes next.

Step 2   Start at 70. Keep counting by 10s.
The numbers 80 and 90 come next.

Answer   The missing numbers are 70 and 90.

Step-by-Step Practice

Skip count by 10s. Write the missing numbers.

10, 20, _____, 40, _____

ten,   twenty,   thirty,   forty,   fifty

Step 1   Use the chart. Start at 10. Skip count by 10s.

The numbers _____ and _____ come next.

Step 2   Start at _____. Keep counting by 10s.

The numbers _____ and _____ come next.

Answer   The missing numbers are _____ and _____.

Name _____

 Guided Practice

Skip count by 10s. Write the missing numbers.

1 10, _____, 30, _____, 50

2 30, _____, _____, 60, _____

3 _____, 60, _____, 80, _____

Problem-Solving Practice

4 Mrs. Sanchez keeps jars of marbles in her classroom.
There are 10 marbles in each jar.
How many marbles are there in all?

Understand   Underline key words.

Plan   Skip count by 10s.

Solve   Count 10 marbles for each jar.

_____, _____, _____, _____, _____, _____, _____, _____

There are _____ marbles in all.

Check   Use 8 dimes and skip count by 10s.

GO on

▶ Practice on Your Own

Skip count by 10s.
Write each missing number or number name.

5 20, _____, 40, 50, _____

6 _____, 60, _____, _____, 90

7 ten, _____, thirty, _____, _____

8 _____, forty, fifty, _____, _____

---

9 Skip count by 10s.

____10____, _____, _____, _____, _____ apples

---

10 **WRITING IN ▶MATH** Theresa has 9 dimes.
How many cents does she have in all? Explain.

_____

_____

Vocabulary Check Complete.

11 A _____ is equal to 10 cents.

STOP

40 forty

Name _____

Progress Check 4 (Lessons 1-7 and 1-8)

Skip count by 5s. Write the missing numbers.

1 10, _____, _____, 25, 30

2 25, 30, _____, 40, _____

3 _____, 35, _____, 45, _____

1	2	3	4	5	6	7	8	9	10
11	12	13	14	15	16	17	18	19	20
21	22	23	24	25	26	27	28	29	30
31	32	33	34	35	36	37	38	39	40
41	42	43	44	45	46	47	48	49	50

Skip count by 10s. Write the missing numbers.

4 10, _____, _____, _____, 50

5 _____, _____, 80, 90, _____

6 Lisa saves 8 nickels. Zina saves
5 dimes. Who has more cents?
Skip count by 5s and 10s. Explain.

Lisa has _____, _____, _____,

_____, _____, _____, _____,

_____ cents.

Zina has _____, _____, _____,

_____, _____ cents.

Lisa

Zina

_____

_____

_____

Name _____

Skip Home

Skip count to find each girl's path home.
Claire skip counts by 5s.
Brandi skip counts by 10s.

Key
Color Claire's path red.
Color Brandi's path blue.

School

Brandi's Home

0	10	15	30	45	60	100
5	20	35	60	70	80	90
10	30	40	50	55	65	95
15	20	30	55	70	70	85
50	25	60	45	50	55	100
40	30	35	40	30	60	95
55	45	50	75	80	65	70

Claire's Home

Name _____

Review

Vocabulary

Word Bank

count

dime

nickel

Use the Word Bank to complete.

1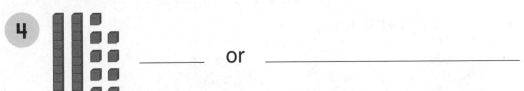
    1 2 3 4 5   ◄┈┈┈┈┈┈┈┈ _____

2   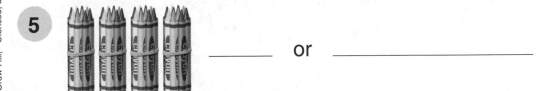   3

   ↑┈┈┈ _____   ↑┈┈┈ _____

▶ **Concepts**

Count. Write the number and the number name.

4   _____ or _____

5   _____ or _____

6   Draw ☐ to show the number 13.

GO on

Write the missing numbers.

7

91		93	94	95	96		98	99	100

8

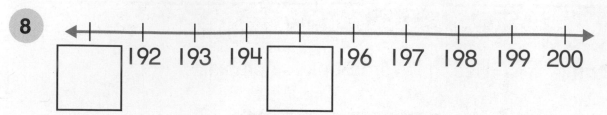

☐   192   193   194   ☐   196   197   198   199   200

---

9 Write the number and the number name.

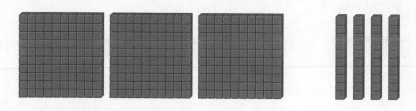

_____   _____   _____

_____   _____   _____

Skip count. Write the missing numbers.

10

_____, _____, _____, _____ puppies

11

_____, _____, _____, _____, _____ fingers

12 _____, 60, 70, _____, _____

STOP

Chapter Test

Count. Write the number and the number name.

1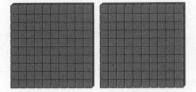

_____ or _____ craft sticks

2

_____   _____   _____

_____   _____   _____

---

3 Draw 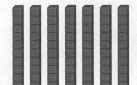 to show the number twenty-three.

Write the missing numbers.

4

| 71 | 72 | | 74 | 75 | 76 | 77 | 78 | 79 | |

5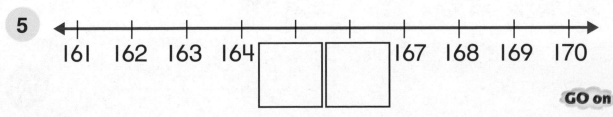

161 162 163 164 ☐ ☐ 167 168 169 170

6 Who is Correct?

Vanessa and Kyung count the cars.

I counted each car. There are 17 cars.

Vanessa

I skip counted by 2s. There are 18 cars.

Kyung

Circle the correct answer. Explain.

_____

_____

Skip count. Write the missing numbers.

7 6, 8, _____, 12, _____

8 _____, 50, 60, _____, _____

9 Greg has saved 7 nickels. How many cents has he saved?

Greg has saved _____ cents.

STOP

Name _____

Test Practice

Listen as your teacher reads each problem.
Choose the correct answer.

1 Which number tells how many baseballs?

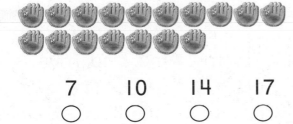

2   10   20   30
○   ○   ○   ○

2 Which number tells how many baseball gloves?

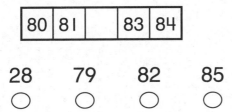

7   10   14   17
○   ○   ○   ○

3 Which number is missing?

80	81		83	84

28   79   82   85
○   ○   ○   ○

4 Which number is missing?

56	57		59	60

55   58   61   85
○   ○   ○   ○

5 Which number tells how many trumpets?

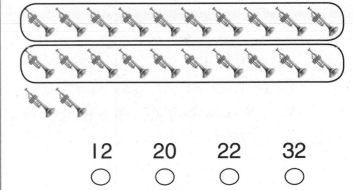

12   20   22   32
○   ○   ○   ○

6 Which number is missing?

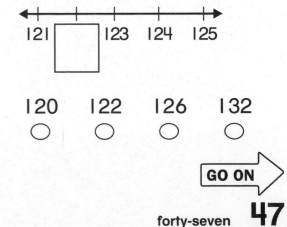

121   123   124   125

120   122   126   132
○   ○   ○   ○

GO ON

7 Which number is modeled?

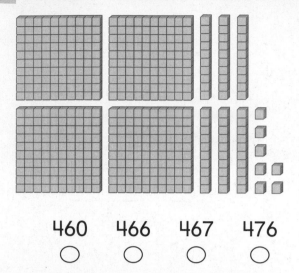

| 460 | 466 | 467 | 476 |
| ○ | ○ | ○ | ○ |

8 Which number is missing?

4, 6, ____, 10, 12

| 2 | 7 | 8 | 9 |
| ○ | ○ | ○ | ○ |

9 Kayla skip counts by 5s. She counts 30, 35, 40, 50. What number does Kayla forget?

30, 35, 40, ____, 50

| 25 | 42 | 45 | 55 |
| ○ | ○ | ○ | ○ |

10 Juan recycles bottles. He receives 5¢ for each bottle. How much will Juan receive for 5 bottles?

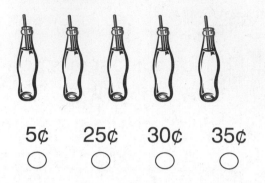

| 5¢ | 25¢ | 30¢ | 35¢ |
| ○ | ○ | ○ | ○ |

11 Which number is missing?

60, 70, 80, 90, ____

| 50 | 95 | 100 | 110 |
| ○ | ○ | ○ | ○ |

12 Tommy, Ama, and Taye catch butterflies. They each catch 10 butterflies. How many butterflies do they catch in all?

| 15 | 20 | 30 | 40 |
| ○ | ○ | ○ | ○ |

STOP

48 forty-eight

Home Connection

English

Spanish

Dear Family,

Today our class started **Chapter 2, Place Value.** In this chapter, I will learn to model numbers and write numbers in word form. I will also learn how to represent numbers in different ways and how to round numbers.

Love, _____

Estimada familia,

Hoy en clase empezó el **Capítulo 2, Valor posicional.** En este capítulo aprenderé a trabajar con números y a escribirlos con palabras. También aprenderé a representar números de diferentes maneras y cómo redondearlos.

Cariños, _____

Help at Home

You can work with your child to model numbers and represent them in different ways. Give them a group of small items, such as beans. Ask them to sort the beans into piles of ten. Ask how many piles of ten they have and how many are left.

Math Online Take the chapter Get Ready quiz at macmillanmh.com.

Ayude en casa

Usted puede trabajar con su niño para que hagan ejemplos de números y los representen de diferentes maneras. Déle un grupo de objetos pequeños, como frijoles. Pídale que agrupe los frijoles en montones de diez. Pregúntele cuántos montones tiene y cuántos le sobraron.

Name _____

Get Ready

Count. Write the number.

1

_____

2

_____

---

3 Compare. Circle the set that has **less**.

4 Compare. Circle the set that has **more**.

Write each number name.

5 3 _____   **6** 11 _____   **7** 15 _____

---

8 How many candles are

on the cake? _____

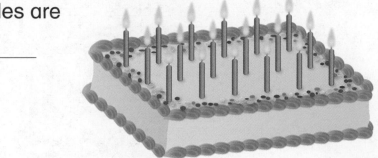

Name _____

Model Numbers 1 to 20 by Ones and Tens

Key Concept

There are thirteen beans.

How many groups of ten can you make? **I group**

How many beans are left? **3 beans**

I ten, 3 ones

$10 + 3 = 13$

Number	Number Name
11	eleven
12	twelve
13	thirteen
14	fourteen
15	fifteen
16	sixteen
17	seventeen
18	eighteen
19	nineteen
20	twenty

Vocabulary

tens place   in 13, the 1 is in the tens place

ones place   in 13, the 3 is in the ones place

13

You can use a place-value chart to show a number. In 5, the 0 is in the tens place and the 5 is in the ones place.

tens	ones
0	5

GO on

Example

There are fifteen leaves.
How many tens and ones are there?

Step 1   Count and circle 10 leaves.
Step 2   Count. There are 5 leaves left.
Step 3   There is 1 ten. There are 5 ones.
Step 4   Fill in the chart.

Answer

tens	ones
1	5

Step-by-Step Practice

There are twelve butterflies.
How many tens and ones are there?

Step 1   Count and circle __10__ butterflies.

Step 2   Count. There are _____ butterflies left.

Step 3   There is _____ ten. There are _____ ones.
Step 4   Fill in the chart.

Answer

tens	ones

Guided Practice

1 Count. Then fill in the chart.
   There are sixteen socks in a drawer.

tens	ones

Problem-Solving Practice

2 Daniel has fourteen marbles.
   How many tens and ones does he have?

Understand   Underline key words.

Plan   Draw a picture.

Solve

Count how many tens and ones.

Daniel has _____ ten and _____ ones.

Check   Did you draw fourteen marbles?
Circle 10 marbles. Count the marbles left.

GO on

▶ Practice on Your Own

Count. Then fill in the chart.

3 Jorge has eleven toy cars.

tens	ones

_____ ten _____ one

4 Marsha has seventeen toy boats.

tens	ones

_____ ten _____ ones

Write the number. Then fill in the chart.

5 four _____

tens	ones

6 nineteen _____

tens	ones

7 **WRITING IN** ▶**MATH** Lilly has fifteen toy trucks.
How many tens and ones does she have? Explain.

_____

_____

Vocabulary Check Complete.

8 Look at the number 18.

The digit 8 is in the _____ place.

STOP

Name _____

Model Numbers 1 to 50 by Ones and Tens

Key Concept

You can use base-ten blocks to show a number as **tens** and **ones**.

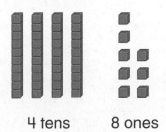

4 tens   8 ones

You can use a **place-value chart** to show tens and ones.

tens	ones
4	8

Vocabulary

tens 23
This number has 2 tens.

ones 23
This number has 3 ones.

place-value chart a chart that shows the place value of digits in a number

tens	ones
2	3

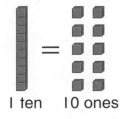

I ten   10 ones

10 ones cubes equal I tens rod

GO on

Example

Count the tens and ones. Then fill in the chart.

Step 1   Count the tens rods.
There are 2 tens rods.

Step 2   Count the ones cubes left.
There are 7 ones cubes left.

Step 3   There are 2 tens.
There are 7 ones.

Answer

tens	ones
2	7

Step-by-Step Practice

Count the tens and ones. Then fill in the chart.

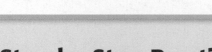

Step 1   Count the tens rods.

There are ___3___ tens rods.

Step 2   Count the ones cubes left.

There are ___8___ ones cubes left.

Step 3   There are _____ tens.

There are _____ ones.

Answer

tens	ones

Name _____

Guided Practice

1   Count the tens and ones. Then fill in the chart.

_____ tens   _____ ones

tens	ones

Problem-Solving Practice

2   Maria has 4 dimes and 2 pennies.
Each dime equals 10 cents. Each penny equals 1 cent.
Count to find how many Maria has.

Understand   Underline key words.

Plan   Act it out.

Solve   Show 4 dimes in the tens box.
Show 2 pennies in the ones box.

tens	ones

_____   _____

Maria has _____ cents.

Check   Skip count by tens for each dime.
Count on one more for each penny.

GO on

▶ Practice on Your Own

Count the tens and ones. Then fill in the chart.

3

tens	ones

_____ tens _____ ones

4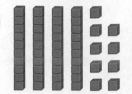

tens	ones

_____ tens _____ ones

5

tens	ones

_____ tens _____ ones

6

tens	ones

_____ tens _____ ones

7 Look at the number 24.
How many ones are in the number 24? _____

8 Look at the number 8.
How many tens are in the number 8? _____

9 **WRITING IN ▶MATH** Matthew has 3 dimes and
2 pennies. How much money does he have? Explain.

_____

_____

Vocabulary Check Complete.

10 A chart that shows place value is a

_____ .

Progress Check 1 (Lessons 2-1 and 2-2)

Count. Then fill in the chart.

1 Shim collected twelve acorns.

tens	ones

_____ ten _____ ones

2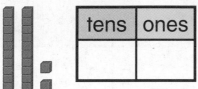

tens	ones

_____ tens _____ ones

3

tens	ones

_____ tens _____ ones

Write the number. Then fill in the chart.

4 eighteen _____

tens	ones

5 five _____

tens	ones

6 twenty _____

tens	ones

7 A number has 2 ones.
It has 2 more tens than ones.
What is the number?

tens	ones

Name _____

«« Replay **Treasure Hunt**

Help Trent find the treasure.
Follow each clue. Draw a line along your path.

1 Start at 1 ten.	2 Go to 1 ten and 3 ones.	3 Go to 2 tens.
4 Go to 3 tens and 5 ones.	5 Go to 4 tens.	6 Go to 2 tens and 2 ones.
7 Go to 4 tens and 6 ones.	8 Go to 2 tens and 8 ones.	9 Finish at 5 tens.

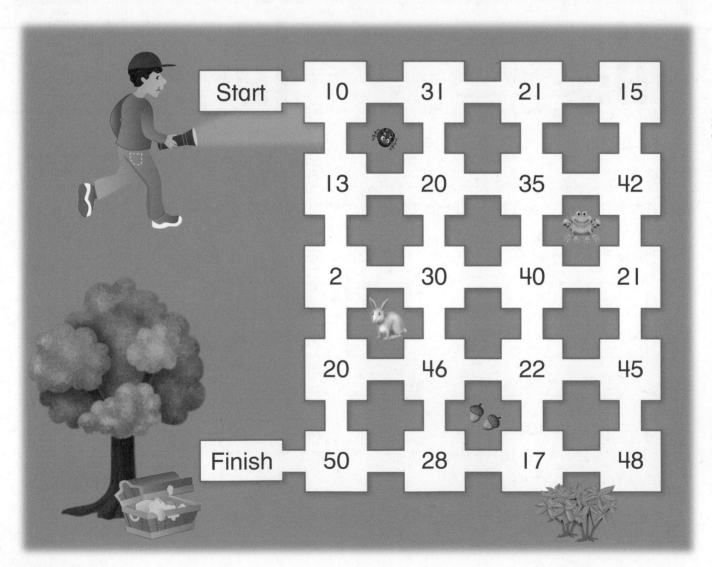

Start | 10 | 31 | 21 | 15

13 | 20 | 35 | 42

2 | 30 | 40 | 21

20 | 46 | 22 | 45

Finish | 50 | 28 | 17 | 48

Name _____

Numbers 1 to 100

Key Concept

You can count by 10s to tell how many.

Look at the crayons. How many crayons are there?

10 ten	20 twenty	30 thirty	40 forty	50 fifty

60 sixty	70 seventy	80 eighty	90 ninety	100 one hundred

There are 100 crayons.

There are one hundred crayons.

Vocabulary

ones 23
This number has 3 ones.

tens 23
This number has 2 tens.

You can also use base-ten blocks to count by tens.

GO on

Example

Count the juice boxes. How many are there?

Step 1   Skip count the groups of ten.
10, 20, 30, 40, 50

Step 2   Count on using the single juice boxes.
Start with 51. 51, 52, 53, 54

Step 3   The ending number is 54.

Answer   There are 54 juice boxes.

Step-by-Step Practice

Count the base-ten blocks.
How many are there?

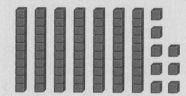

Step 1   Skip count the tens rods.

___10___ , ___20___ , ___30___ , _____ , _____ ,

_____ , _____

Step 2   Count on using the ones cubes.

Start with _____ . _____ , _____ , _____ ,

_____ , _____ , _____ , _____ , _____

Step 3   The ending number is _____ .

Answer   There are _____ base-ten blocks.

Name _____

 Guided Practice

Count. Write the number.

1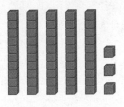

_____ tens _____ ones

2

_____ tens _____ ones

Problem-Solving Practice

3 Kellie has these bowls of cherries.
How many cherries does Kellie have?

Understand   Underline key words.

Plan   Use a model.

Solve   Skip count by 10s.

_____, _____, _____, _____, _____, _____

Kellie has _____ cherries.

Check   Count the cherries.
Cross out each cherry as you count.

GO on

▶ Practice on Your Own

Count. Write the number.

4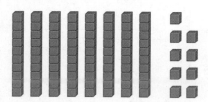

_____ tens _____ ones

5

_____ tens _____ ones

6

_____ _____

7

_____ _____ _____ _____

---

8 ◖ **WRITING IN** ▶**MATH**   Niabi has these crayons.
How many crayons does Niabi have? Explain.

_____

_____

_____

Vocabulary Check   Complete.

9   The number 63 has 3 _____.

STOP

Name _____

Numbers 1 to 500

Key Concept

You can use base-ten blocks to show a number as **hundreds**, tens, and ones.

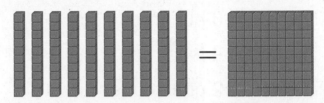

10 tens rods equal 1 hundreds flat

hundreds	tens	ones
1	4	2

1 hundreds flat   4 tens rods   2 ones cubes

100   +   40   +   2   = 142

Vocabulary

hundreds   the numbers 100–999

234
This number has
2 hundreds.

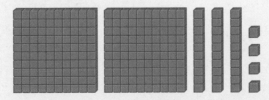

GO on

Example

Count. Then fill in the chart.

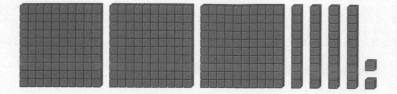

Step 1   Count. There are 3 hundreds flats.

Step 2   Count. There are 4 tens rods.

Step 3   Count. There are 2 ones cubes.

Answer

hundreds	tens	ones
3	4	2

Step-by-Step Practice

Count. Then fill in the chart.

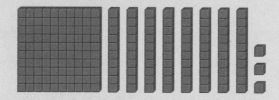

Step 1   Count. There is _____ hundreds flat.

Step 2   Count. There are _____ tens rods.

Step 3   Count. There are _____ ones cubes.

Answer

hundreds	tens	ones

Name _____

 Guided Practice

Count. Then fill in the chart.

1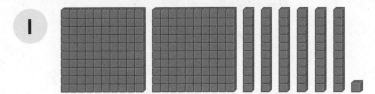

hundreds	tens	ones

2

hundreds	tens	ones

Problem-Solving Practice

3 Mr. Maher's class collected 259 cereal box tops.
How many hundreds, tens, and ones is this?

Understand Underline key words.

Plan Count the hundreds, tens, and ones.

Solve Use a place-value chart.

hundreds	tens	ones

They collected _____ hundreds,

_____ tens, and _____ ones.

Check Use base-ten blocks to act it out.

GO on

▶ Practice on Your Own

Count. Then fill in the chart.

4

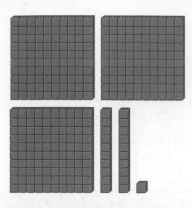

hundreds	tens	ones

5

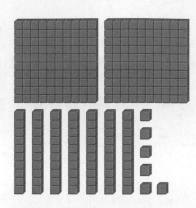

hundreds	tens	ones

6 Look at the number 315. Fill in the blanks.

The 3 is in the _____ place.

The 1 is in the _____ place.

7 **WRITING IN ▶MATH**   Kiah has 408 seashells.
How many hundreds, tens, and ones is 408?
How do you know?

_____

_____

_____

_____

Vocabulary Check   Complete.

8 In the number 365, there are 3 _____.

68   sixty-eight

Name _____

Progress Check 2 (Lessons 2-3 and 2-4)

Count. Then fill in the chart.

1 Basir has eighty-three crayons.

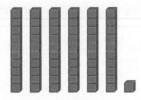

hundreds	tens	ones

2

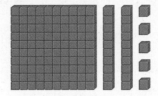

hundreds	tens	ones

Count. Write the number.

3

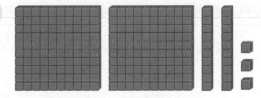

_____ tens _____ one

4

_____ _____ _____

5 Alicia is thinking of a number.
The number has 4 tens and 4 ones.
The number has 1 less hundred than tens.
What is the number?

hundreds	tens	ones

Name _____

«« Replay **Lost Dogs**

Five runners lost their dogs in the park.
Match the pets to their owners.

Draw a line from each runner to the correct dog.

1	My owner has 1 hundred, 7 tens, and 6 ones. My owner's number is _____.

2	My owner has 1 hundred, 3 tens, and 6 ones. My owner's number is _____.

3	My owner has 2 hundreds, 4 tens, and 9 ones. My owner's number is _____.

4	My owner has 3 hundreds, 6 tens, and 1 one. My owner's number is _____.

5	My owner has 4 hundreds, 8 tens, and 2 ones. My owner's number is _____.

70 seventy

Name _____

Numbers 1 to 1,000

Key Concept

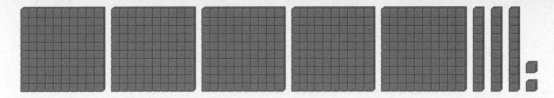

There are 5 hundreds flats.
Count by 100s. 100, 200, 300, 400, 500

There are 3 tens rods.
Count by 10s. 10, 20, 30

There are 2 ones cubes.
Count by 1s. 1, 2

      5 hundreds   3 tens   2 ones
      500   + 30   + 2   =   532

The digits 5, 3, and 2 are used to make the number 532.

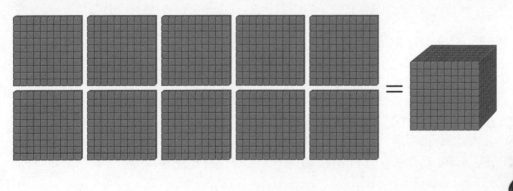

10 hundreds flats
are equal to 1,000.

GO on

Count. Then fill in the chart.

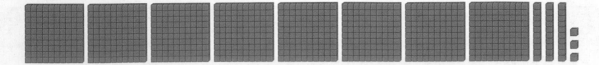

Step 1   Count the hundreds. There are 8 hundreds.

Step 2   Count the tens. There are 3 tens.

Step 3   Count the ones. There are 3 ones.

Answer

hundreds	tens	ones
8	3	3

Step-by-Step Practice

Count. Then fill in the chart.

Step 1   Count the hundreds.

There are ___7___ hundreds.

Step 2   Count the tens. There are _____ tens.

Step 3   Count the ones. There are _____ ones.

Answer

hundreds	tens	ones

Name _____

▶ Guided Practice

Count. Then fill in the chart.

1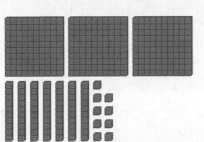

hundreds	tens	ones

2

hundreds	tens	ones

Problem-Solving Practice

3 Rosa has 2 one-dollar bills, 5 dimes, and 3 pennies.

How many cents does Rosa have?

Understand Underline key words.

Plan Use a model.

Solve For each dollar count 100. _____

For each dime count 10. _____

For each penny count 1. _____

Rosa has _____ cents.

Check Use a place-value chart.

GO on

▶ Practice on Your Own

Count. Then fill in the chart.

4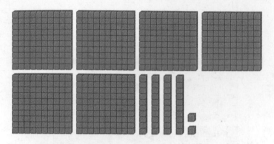

hundreds	tens	ones

5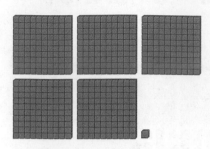

hundreds	tens	ones

6 Look at the number 976. Fill in the blanks.

The 9 is in the _____ place.

The 7 is in the _____ place.

7 **WRITING IN ▶MATH**   Rashida has 7 one-dollar bills and 3 pennies. How many cents does Rashida have? Explain.

_____

_____

_____

Vocabulary Check   Complete.

8 In the number 867, the digit 7 is in the

_____ place.

STOP

Name _____

Short Word Form

Key Concept

Count the tens and ones.

There are 3 tens.
There are 6 ones.

How many tens and ones are used to show
this number? 3 tens + 6 ones = 36

This is how the number is written in **short word form**.

3 tens 6 ones

Vocabulary

hundreds   the numbers 100–999

234

This number has 2 hundreds.

short word form   a way of writing numbers
that uses digits and place value words

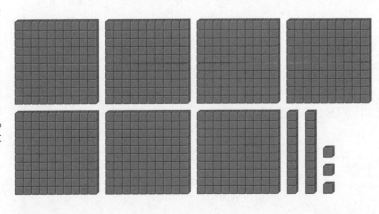

You can also write hundreds
in short word form.

7 hundreds 2 tens 3 ones

GO on

Example

Write **two hundred forty-six** in short word form.

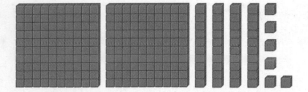

Step 1	Count the hundreds flats. There are 2 hundreds.
Step 2	Count the tens rods. There are 4 tens.
Step 3	Count the ones cubes. There are 6 ones.
Answer	2 hundreds 4 tens 6 ones

Step-by-Step Practice

Write three hundred fourteen in short word form.

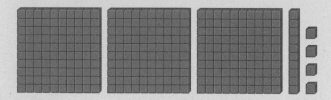

Step 1	Count the hundreds flats. There are ___3___ hundreds.
Step 2	Count the tens rods. There is _____ ten.
Step 3	Count the ones cubes. There are _____ ones.
Answer	_____ hundreds _____ ten _____ ones

I Write the number in short word form.

There are four hundred sixty-seven children in a school.

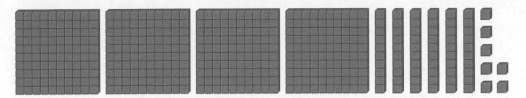

_____ hundreds _____ tens _____ ones

Problem-Solving Practice

2 Obike has 62 books.
Each box will hold 10 books.
How many boxes can Obike fill?
How many books will be left?

Understand Underline key words.

Plan Draw a picture.

Solve

Count the boxes. Count the books left.

Obike can fill _____ boxes.

There are _____ books left.

Check Skip count by 10s. Count on by 1s.

GO on

▶ Practice on Your Own

Write the short word form of each number.

3   17 = _____ ten _____ ones

4   43 = _____ tens _____ ones

5   534 = _____ hundreds _____ tens _____ ones

6   Samantha picked twenty-three strawberries.

_____ tens _____ ones

7   **WRITING IN ▶ MATH**   The Douglas Farm picked six hundred forty-two tomatoes. What is the short word form of this number? Explain.

_____

_____

_____

Vocabulary Check   Complete.

8   In the number 456, 4 is in the _____ place.

STOP

Progress Check 3 (Lessons 2-5 and 2-6)

Count. Then fill in the chart.

1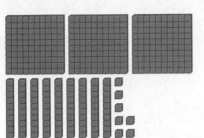

hundreds	tens	ones

2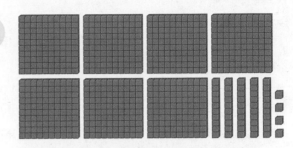

hundreds	tens	ones

Write the short word form of each number.

3 A car repair shop ordered
four hundred twenty-four tires.

_____ hundreds _____ tens _____ ones

4 37 = _____ tens _____ ones

5 126 = _____ hundred _____ tens _____ ones

---

6 Gabrielle is thinking of a number.
The number has 5 tens and 3 hundreds.
The ones digit is the same as the
hundreds digit. What is the number?

hundreds	tens	ones

Name _____

Greater Number Game

How to Play

1 Write your name and your partner's name on a sheet of paper.

2 Take turns tossing three number cubes. Use the digits to make a number.

3 Write the number.

4 Compare the number made by each player. Circle the greater number.

5 The player with the greater number gets 1 point. The first player to get 10 points wins.

Materials
- 3 number cubes
- paper and pencil

Elena	Kwasi
(621)	543
332	(431)

Name _____

Writing Numbers

Key Concept

You can show a number in different ways.
Here are some ways to show 36.

Short Word Form
3 tens 6 ones

Using Digits
36

Using Addition
30 + 6

Using Words
thirty-six

Vocabulary

place value   the value given to a digit by its
place in a number

365

3 is in the hundreds place.
6 is in the tens place.
5 is in the ones place.

hundreds	tens	ones
3	6	5

GO on

Example

There are 45 goldfish in a tank.
Write this number in different ways.

Step 1   The digit 4 is in the tens place.

Step 2   The digit 5 is in the ones place.

Step 3   Write the number in short word form.

Step 4   Write the number using addition.

Answer

Short Word Form
4 tens 5 ones

Using Addition
40 + 5

Step-by-Step Practice

There are 72 starfish in an aquarium.
Write this number in different ways.

Step 1   The digit _____ is in the tens place.

Step 2   The digit _____ is in the ones place.

Step 3   Write the number in short word form.

Step 4   Write the number using addition.

Answer

Short Word Form
_____ tens _____ ones

Using Addition
_____ + _____

Name _____

Write each number in different ways.

1 There are twenty-six buttons in a basket.

Using Digits	Using Addition
_____	_____ + _____

2 Stephanie has 35 marbles.

Short Word Form	Using Words
_____ tens _____ ones	_____

Problem-Solving Practice

3 Akira has 2 pages with 10 kitten stickers each.
He has 1 page with 3 kitten stickers.
How many kitten stickers does Akira have in all?

Understand   Underline key words.

Plan   Use base-ten blocks.

Solve   Show _____ tens rods. Show _____ ones cubes.

     Akira has _____ kitten stickers.

Check   Use a place-value chart.

GO on

▶ Practice on Your Own

Write each number in different ways.

4 The band has 16 flutes.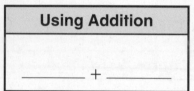

Using Words	Using Addition	Short Word Form
_____	_____ + _____	_____ ten _____ ones

5 The band has thirty-one trumpets.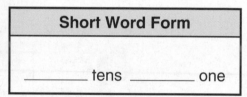

Using Digits	Using Addition	Short Word Form
_____	_____ + _____	_____ tens _____ one

Fill in the blanks.

6 What number can be written as 50 + 3? _____

7 What number has 2 tens and 5 ones? _____

8 ✏ **WRITING IN ▶MATH** Paul has 7 dimes and 2 pennies. How many cents does Paul have? How do you know?

_____

_____

Vocabulary Check Complete.

9 The value given to a digit by its place in a number

is _____.

Name _____

Round Using a Number Line

Key Concept

The number 13 is between 10 and 20.
Round 13 to the nearest 10.

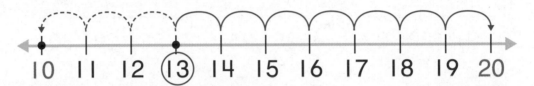

The number 13 is 3 away from 10.
The number 13 is 7 away from 20.
The number 13 is closer to 10.
13 rounds down to 10.

Vocabulary

round to change the value of a number to one that is easier to work with

When a number has a 5 in the ones place, round up.

number line a line with number labels

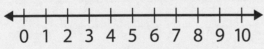

There are 5 ones in 35.
35 rounds up to 40.

GO on

Copyright © Macmillan/McGraw-Hill, • Glencoe, a division of The McGraw-Hill Companies, Inc.

Example

There are 24 students on a bus. About how many students are on the bus? Round to the nearest ten.

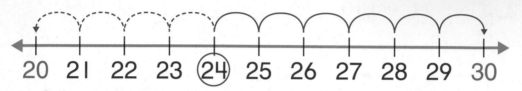

20 21 22 23 (24) 25 26 27 28 29 30

Step 1   Circle 24. The nearest ten to the left is 20.
Step 2   The nearest ten to the right is 30.
Step 3   The number 24 is closer to 20.

Answer   There are about 20 students on the bus.

Step-by-Step Practice

There are 47 students on the playground. About how many students are on the playground? Round to the nearest ten.

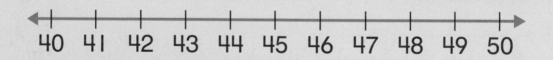

40 41 42 43 44 45 46 47 48 49 50

Step 1   Circle 47. The nearest ten to the left is __40__.

Step 2   The nearest ten to the right is _____.

Step 3   The number 47 is closer to _____.

Answer   There are about _____ students on the playground.

86   eighty-six

Name _____

▶ Guided Practice

Round each number to the nearest ten.

1 54 _____

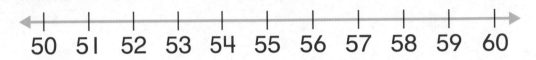

50 51 52 53 54 55 56 57 58 59 60

2 82 _____

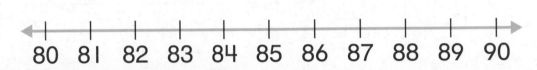

80 81 82 83 84 85 86 87 88 89 90

Problem-Solving Practice

3 Miguel has 95 sand dollars.
About how many sand dollars does he have?
Round to the nearest ten.

Understand   Underline key words.

Plan   Use a number line.

Solve   Circle the number 95.
Round 95 to the nearest ten.

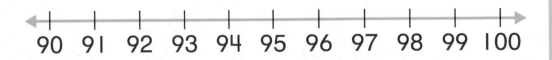

90 91 92 93 94 95 96 97 98 99 100

Miguel has about _____ sand dollars.

Check   If the ones digit is greater than or equal to 5,
round up. Does your answer follow this rule?

GO on

▶ Practice on Your Own

Round each number to the nearest ten.

4 13 _____

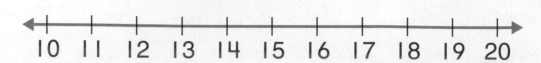

5 35 _____

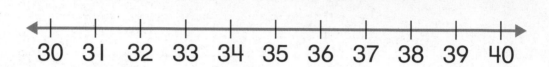

6 42 _____

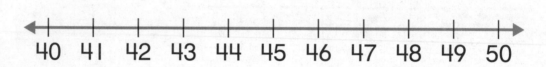

7 71 _____

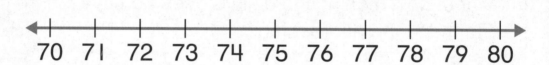

---

8 **WRITING IN ▶MATH** The distance from Ahmed's house to the beach is exactly 85 miles. Round 85 to the nearest ten. Explain.

_____

_____

_____

Vocabulary Check Complete.

9 A _____ is a line with number labels.

STOP

Progress Check 4 (Lessons 2-7 and 2-8)

Write each number in different ways.

1

Using Addition
_____ + _____

Short Word Form
_____ tens _____ ones

2

Using Digits
_____

Using Words
_____

Round each number to the nearest ten.

3 24 _____

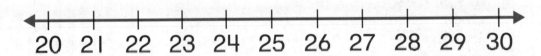

4 78 _____

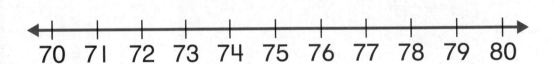

5 92 _____

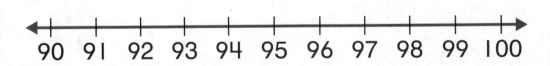

6 Ogima says he can show 39 with 2 tens rods
and 19 ones cubes. Is he correct? Explain.

_____

Name _____

«« Replay

Haunted Party

Match each ghost to the correct letter. Use the letters to complete the riddle.

What kind of tie does a ghost wear to a haunted party?

O	
A	40 + 6
E	
I	70 + 7
B	
T	
O	80 + 5

_____ _____ _____ _____   _____ _____ _____
46   32   58   85   24   77   69

Name _____

Review

Word Bank

hundreds place

number line

ones place

tens place

Use the Word Bank to complete.

1.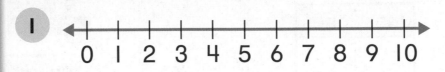

   _____

2. 327

   _____

3. 12

   _____

4. 864

   _____

 **Concepts**

Count. Then fill in the chart.

5.

tens	ones

6.

tens	ones

GO on

Count. Then fill in the chart.

7

tens	ones

8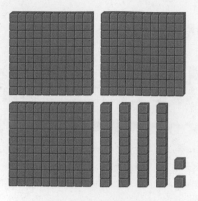

hundreds	tens	ones

Write the short word form of each number.

9

_____2_____ tens _____ ones

10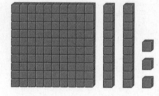

_____ hundred

_____ tens _____ ones

Write each number as tens and ones.

11 47 _____4_____ tens _____ ones

12 28 _____ tens _____ ones

Round each number to the nearest ten.

13 41 _____40_____

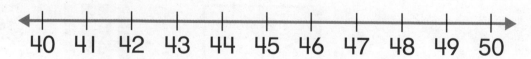

40 41 42 43 44 45 46 47 48 49 50

14 87 _____

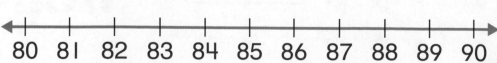

80 81 82 83 84 85 86 87 88 89 90

STOP

92 ninety-two

Name _____

Chapter Test

Count. Then fill in the chart.

1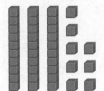

tens	ones

_____ tens _____ ones

2

tens	ones

_____ tens _____ one

Count. Write the number.

3

_____

4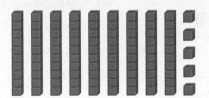

_____

Write the short word form of each number.

5

_____ hundred _____ tens _____ ones

6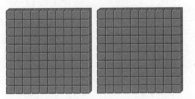

_____ hundreds _____ ten _____ ones

7 56 _____

8 63 _____

9 Who is Correct?

Shani and Gustavo round 43 to the nearest ten.

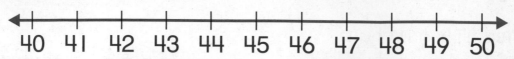

40 41 42 43 44 45 46 47 48 49 50

The number 43 rounds up to 50.

Shani

Gustavo

The number 43 rounds down to 40.

Circle the correct answer. Explain.

_____

_____

10 Round the number to the nearest ten.

There are 73 passengers on a ship. _____

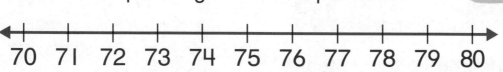

70 71 72 73 74 75 76 77 78 79 80

11 Look at the model.
Explain how to write the number in different ways.

_____

_____

STOP

94 ninety-four

Name _____

Test Practice

Listen as your teacher reads each problem.
Choose the correct answer.

1 Which is the number name for 16?

   six   sixteen   thirty   sixty
   ○   ○   ○   ○

2 Which number has 6 tens and 8 ones?

   14   28   68   86
   ○   ○   ○   ○

3 How many hearts are there?

   12   15   17   20
   ○   ○   ○   ○

4 Which number has 3 tens and 7 ones?

   10   30   37   73
   ○   ○   ○   ○

5 Which number has 7 tens and no ones?

   7   17   57   70
   ○   ○   ○   ○

6 Which number is shown by the place-value chart?

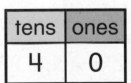

tens	ones
4	0

   four   forty   fifty   sixty
   ○   ○   ○   ○

GO ON →

7 Which is another way to write twenty-three?

2 + 3 ○   20 + 3 ○

30 + 2 ○   40 + 3 ○

8 There are 27 birds in a tree. About how many birds are in the tree?

20 21 22 23 24 25 26 27 28 29 30

2 ○   20 ○   30 ○   40 ○

9 Which is the number for 800 + 20 + 9?

82 ○   89 ○   809 ○   829 ○

10 There are 18 soccer players on a field. About how many players are on the field?

10 ○   20 ○   30 ○   80 ○

11 Which is the place value of the 3 in 53?

ones ○   hundreds ○

tens ○   fives ○

12 Which is another way to show 526?

○ 50 + 6
○ 50 + 2 + 6
○ 500 + 6
○ 500 + 20 + 6

13 There are fifty-two marbles in a bag. How many tens and ones are there?

○ 1 ten 1 one
○ 2 tens 5 ones
○ 5 tens 0 ones
○ 5 tens 2 ones

STOP

Home Connection

English   **Spanish**

Dear Family,
Today our class started **Chapter 3, Compare and Order Whole Numbers.** In this chapter, I will learn to compare whole numbers up to 1,000. I will also learn to use place value to order three whole numbers up to 500.

Love, _____

Estimada familia,
Hoy en clase comenzamos el **Capítulo 3, Comparar y ordenar números enteros.** En este capítulo aprenderé a comparar números enteros hasta 1,000. También aprenderé a usar el valor posicional para ordenar tres números enteros hasta 500.

Cariños, _____

Help at Home

Show your child three different containers varying in size. Help your child find the number of ounces on each container. Then, have your child compare and order the number of ounces.

Math Online > Take the chapter Get Ready quiz at macmillanmh.com.

Ayude en casa

Muestre a su niño tres recipientes de diferentes tamaños. Ayude a su niño a encontrar el número de onzas de cada recipiente. Luego, pídale que compare y ordene el número de onzas.

Name _____

Get Ready

Count. Write the number. Write the number name.

1 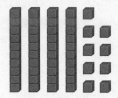 _____ _____

2 

_____ _____

Write the missing numbers.

3

	14	15	16	17	18	19	20		22

4

51	52	53		55	56		58	59	60

5 Look at the number 985.

How many hundreds are there? _____ hundreds

How many tens are there? _____ tens

How many ones are there? _____ ones

Write each number in short word form.

6 82 _____

7 736 _____

Name _____

Compare Numbers 0 to 50

Key Concept

You can use a number line to compare numbers.

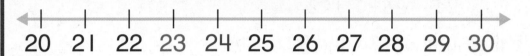

20 21 22 23 24 25 26 27 28 29 30

The lesser number is on the left.

23 is **less than** 30.
23   <   30

Look at the symbols < and >.
The symbol points to the lesser number.

29 is **greater than** 24.
29   >   24

Vocabulary

less than <

4 < 7
4 is less than 7.

greater than >

7 > 2
7 is greater than 2.

equal to =

6 = 6
6 is equal to 6.

Example

Compare. Write >, <, or =.

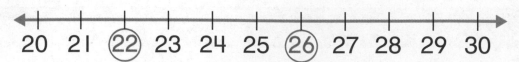

26 ◯ 22

Step 1	Circle the numbers 26 and 22 on the number line.
Step 2	The number 22 is to the left of the number 26. 26 is greater than 22.
Step 3	The symbol should point to 22.

Answer   26 ⟨ > ⟩ 22

Step-by-Step Practice

Compare. Write >, <, or =.

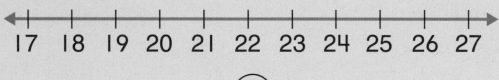

19 ◯ 27

Step 1   Circle the numbers _____ and _____ on the number line.

Step 2   The number _____ is to the left of the

number _____. 19 is _____ than 27.

Step 3   The symbol should point to _____.

Answer   19 ◯ 27

Name _____

 Guided Practice

Compare. Write >, <, or =.

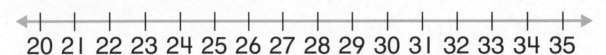

20 21 22 23 24 25 26 27 28 29 30 31 32 33 34 35

I 24 ◯ 20   **2** 23 ◯ 35   **3** 30 ◯ 30

Problem-Solving Practice

4 Dalila scored 25 points in a basketball game. Gamal scored 28 points. Who scored more points?

Understand   Underline key words.

Plan   Use a number line.

Solve   Find the numbers _____ and _____ on the number line above. The greater number is on the right.

The number _____ is on the right.

25 ◯ 28

_____ scored more points.

Check   Use base-ten blocks to model the numbers.

GO on

▶ Practice on Your Own

Compare. Write >, <, or =.

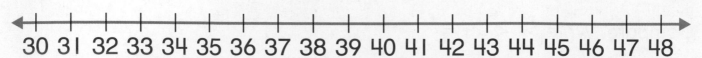

30 31 32 33 34 35 36 37 38 39 40 41 42 43 44 45 46 47 48

5 45 ◯ 45   **6** 36 ◯ 43   **7** 46 ◯ 36

8 42 ◯ 48   **9** 33 ◯ 31   **10** 41 ◯ 47

11 30 ◯ 35   **12** 40 ◯ 40   **13** 44 ◯ 42

14 38 ◯ 39   **15** 37 ◯ 34   **16** 43 ◯ 43

---

17 **WRITING IN ▶ MATH**   Michael has 39 trading cards. Alonso has 44 trading cards. Who has fewer trading cards? Explain.

_____

_____

_____

Vocabulary Check   Complete.

18 There are 6  in Doug's pond.

There are 3 ![fish] in Jackie's pond.   6 ◯ 3

The number 6 is _____ 3.

Name _____

Compare Numbers 0 to 100

Key Concept

You can use base-ten blocks to compare numbers.

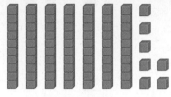

77

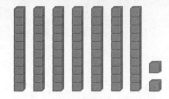

72

Compare tens.	7 tens	=	7 tens
Compare ones.	7 ones	>	2 ones
Which number is greater?	77	>	72

77 is **greater than** 72.

Vocabulary

less than <

4 < 7
4 is less than 7.

greater than >

7 > 2
7 is greater than 2.

equal to =

6 = 6
6 is equal to 6.

Tenisha compares 30 and 44.
3 tens is less than 4 tens.
So, 30 is less than 44. 30 < 44

Compare. Write >, <, or =.

   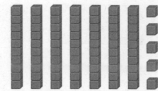

67   75

Step 1   Compare the tens.

           6 tens < 7 tens

Step 2   67 has fewer tens than 75.

Answer   67 (<) 75

Step-by-Step Practice

Compare. Write >, <, or =.

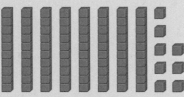 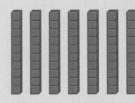

88   81

Step 1   Compare the _____.

       _____ tens ◯ _____ tens

Step 2   Compare the _____.

       _____ ones ◯ _____ one

Step 3   88 has _____ ones than 81.

Answer   88 ◯ 81

Name _____

 Guided Practice

Compare. Write >, <, or =.

1

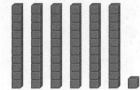

61  45

2

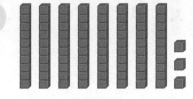

83 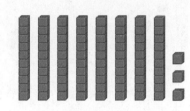 83

Problem-Solving Practice

3 Juan read 62 pages of his book on Monday.
   He read 52 pages of his book on Tuesday.
   Which day did Juan read fewer pages?

   Understand   Underline key words.

   Plan   Use base-ten blocks.

   Solve   Model 62 and 52. Compare.   62 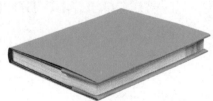 52

   Juan read fewer pages on _____.

   Check   Use a hundred chart to compare the numbers.

GO on

▶ Practice on Your Own

Compare. Write >, <, or =.

4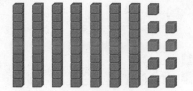

79 ◯ 79

5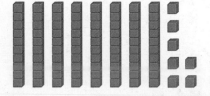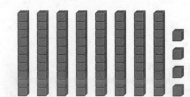

87 ◯ 84

6 65 ◯ 58   **7** 68 ◯ 98   **8** 71 ◯ 73

---

9 **WRITING IN ▶MATH**   Rafi bought a pack of 76 markers.
Kym bought a pack of 78 markers. Who has more markers?
How do you know?

_____

_____

Vocabulary Check   Complete.

10 Gina has . Pablo has .

64 ◯ 64

The number 64 is _____ 64.

Name _____

Progress Check 1 (Lessons 3-1 and 3-2)

Compare. Write >, <, or =.

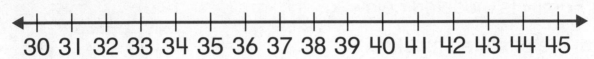

30 31 32 33 34 35 36 37 38 39 40 41 42 43 44 45

1 36 ◯ 40   **2** 44 ◯ 37   **3** 32 ◯ 42

4

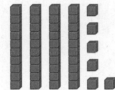

46   ◯   46

5

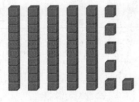 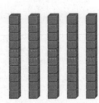

56   ◯   50

6 64 ◯ 66   **7** 95 ◯ 95   **8** 74 ◯ 73

9 Brady solved 12 math problems on Thursday. He solved 21 math problems on Friday. Which day did Brady solve more math problems? Explain.

_____

Name _____

The height of each member of the
Davis Family is modeled below.

Look at the base-ten blocks.
Count the blocks and write the number.

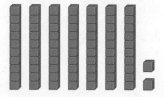

Dad is _____
inches tall.

Grandmom is
_____ inches tall.

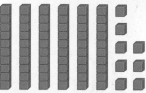

Mom is _____
inches tall.

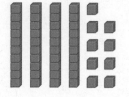

Taye is _____
inches tall.

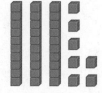

Ama is _____
inches tall.

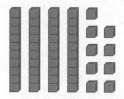

Jamil is _____
inches tall.

Write a number on each tape measure to make a true statement.
Use each number only once.

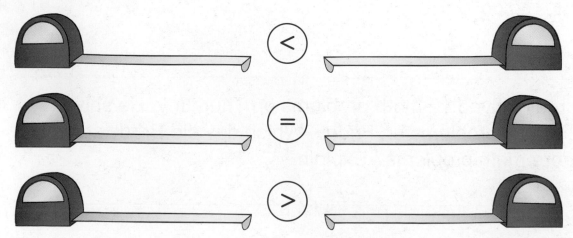

Name _____

Compare Numbers 100 to 500 by Tens and Hundreds

Key Concept

Use place value and base-ten blocks to compare numbers.

 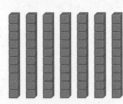

I hundred   7 tens   8 ones = 178

 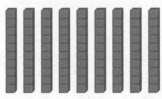

I hundred   9 tens   4 ones = 194

Both numbers have I hundred.

178 has 7 tens. 194 has 9 tens.

7 tens < 9 tens

178 < 194

178 is **less than** 194.

Vocabulary

less than <   4 < 7
               4 is less than 7.

greater than >   7 > 2
               7 is greater than 2.

equal to =   6 = 6
               6 is equal to 6.

GO on

Example

Compare. Write >, <, or =.

142 ◯ 183

Step 1 Write each number in short word form.
142 = 1 hundred 4 tens 2 ones
183 = 1 hundred 8 tens 3 ones

Step 2 Compare the hundreds.
1 hundred = 1 hundred
The hundreds are equal.

Step 3 Compare the tens. 4 tens < 8 tens

Answer 142 ⟨ < ⟩ 183

Step-by-Step Practice

Compare. Write >, <, or =.

346 ◯ 324

Step 1 Write each number in short word form.

346 = _____ hundreds _____ tens _____ ones

324 = _____ hundreds _____ tens _____ ones

Step 2 Compare the hundreds.

_____ hundreds ◯ _____ hundreds

The hundreds are _____.

Step 3 Compare the _____. _____ tens ◯ _____ tens

Answer 346 ◯ 324

Name _____

Compare. Write >, <, or =.

1

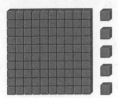   

_____ ◯ _____

2

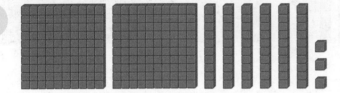  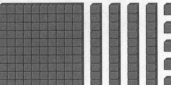

_____ ◯ _____

Problem-Solving Practice

3 Mr. Alvarez drove 296 miles last week. He drove 254 miles this week. Which week did Mr. Alvarez drive more miles?

Understand Underline key words.

Plan Use short word form.

Solve 296 = _____ hundreds _____ tens _____ ones

254 = _____ hundreds _____ tens _____ ones

Compare. 296 ◯ 254

Mr. Alvarez drove more miles _____ week.

Check Model both numbers using base-ten blocks.

▶ Practice on Your Own

Compare. Write >, <, or =.

4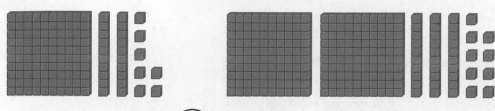

_____ ◯ _____

5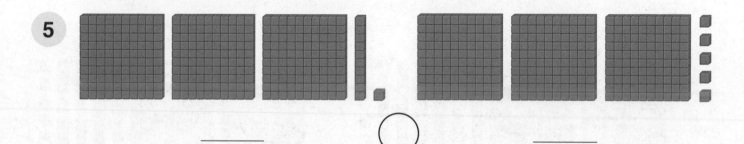

_____ ◯ _____

6 436 = _____ hundreds _____ tens _____ ones

459 = _____ hundreds _____ tens _____ ones

436 ◯ 459

7 286 ◯ 357   **8** 456 ◯ 445   **9** 500 ◯ 500

10 **WRITING IN ▶MATH** A furniture company made 133 sofas in February. They made 152 sofas in March. In which month did they make fewer sofas? Explain.

_____

_____

Vocabulary Check   Complete.

11 342 ◯ 361   4 tens is _____ 6 tens.

STOP

Name _____

Compare Numbers 500 to 1,000 by Tens and Hundreds

Key Concept

Use place value and base-ten blocks to compare numbers.

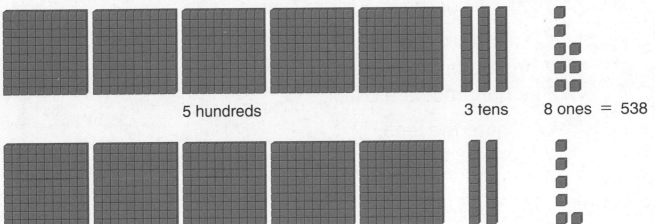

5 hundreds   3 tens   8 ones = 538

5 hundreds   2 tens   6 ones = 526

Both numbers have 5 hundreds.
538 has 3 tens. 526 has 2 tens.
3 tens > 2 tens
538 > 526
538 is **greater than** 526.

Vocabulary

less than <   4 < 7
                 4 is less than 7.

greater than >   7 > 2
                 7 is greater than 2.

equal to =   6 = 6
                 6 is equal to 6.

GO on

Example

Compare. Write >, <, or =.

771 ◯ 787

Step 1   Write each number in short word form.
          771 = 7 hundreds 7 tens 1 one
          787 = 7 hundreds 8 tens 7 ones

Step 2   Compare the hundreds.
          7 hundreds = 7 hundreds
          The hundreds are equal.

Step 3   Compare the tens.
          7 tens < 8 tens

Answer   771 ◯< 787

Step-by-Step Practice

Compare. Write >, <, or =.

654 ◯ 713

Step 1   Write each number in short word form.

          654 = _____ hundreds _____ tens _____ ones

          713 = _____ hundreds _____ ten _____ ones

Step 2   Compare the hundreds.

          _____ hundreds ◯ _____ hundreds

Answer   654 ◯ 713

Name _____

▶ Guided Practice

Compare. Write >, <, or =.

1. 504 = _____ hundreds _____ tens _____ ones

   626 = _____ hundreds _____ tens _____ ones

   504 ◯ 626

2. 827 = _____ hundreds _____ tens _____ ones

   808 = _____ hundreds _____ tens _____ ones

   827 ◯ 808

Problem-Solving Practice

3. In first grade, Mika earned 598 homework points.
   In second grade, he earned 608 homework points.
   In which grade did Mika earn fewer homework points?

   ✓20 points
   1. 4 tens
   2. 9 ones
   3. 2 tens

   Understand Underline key words.

   Plan Use short word form.

   Solve 598 = _____ hundreds _____ tens _____ ones

   608 = _____ hundreds _____ tens _____ ones

   Compare. 598 ◯ 608

   Mika earned fewer points in _____ grade.

   Check Model both numbers using base-ten blocks.

GO on

▶ Practice on Your Own

Compare. Write >, <, or =.

4

_____ hundreds _____ tens _____ one = _____

_____ hundreds _____ tens _____ ones = _____

531 ◯ 544

5 918 = _____ hundreds _____ ten _____ ones

855 = _____ hundreds _____ tens _____ ones

918 ◯ 855

6 724 ◯ 635   **7** 821 ◯ 838   **8** 905 ◯ 905

9 ✏ **WRITING IN** ▶ **MATH**   Janine has 786 video game points. Esi has 774 video game points. Who has more points? Explain.

_____

_____

Vocabulary Check   Complete.

10 791 is _____ 737.   791 ◯ 737   **STOP**

Progress Check 2 (Lessons 3-3 and 3-4)

Compare. Write >, <, or =.

1 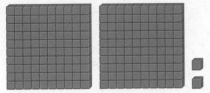 ⬭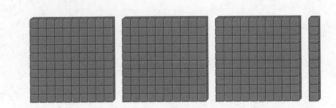

_____ ⬭ _____

2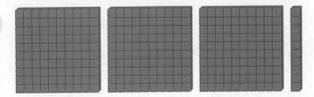

_____ ⬭ _____

3 672 = _____ hundreds _____ tens _____ ones

533 = _____ hundreds _____ tens _____ ones

672 ⬭ 533

4 711 = _____ hundreds _____ ten _____ one

741 = _____ hundreds _____ tens _____ one

711 ⬭ 741

5 376 ⬭ 218 **6** 1,000 ⬭ 1,000 **7** 832 ⬭ 810

8 Shelby made 221 cards in May. She made 236 cards in June. Which month did she make more cards? Explain.

_____

Name _____

Choose the number that makes each statement true.
Use your answers to decide which bridge Jamaal should use.

1 246 < _____   ⬭ 58   ⬭ 158   ⬭ 258

2 377 > _____   ⬭ 331   ⬭ 431   ⬭ 531

3 500 = _____   ⬭ 400   ⬭ 500   ⬭ 600

4 699 > _____   ⬭ 569   ⬭ 769   ⬭ 869

5 924 < _____   ⬭ 787   ⬭ 887   ⬭ 987

Jamaal should use

Bridge _____.

Help me get across the river!

Copyright © Macmillan/McGraw-Hill • Glencoe, a division of The McGraw-Hill Companies, Inc.

Name _____

Compare and Order Numbers to 100

Key Concept

You can use a hundred chart to **compare** and **order** numbers.

Oscar shades the numbers 41, 58, and 63 on a hundred chart.

41	42	43	44	45	46	47	48	49	50
51	52	53	54	55	56	57	58	59	60
61	62	63	64	65	66	67	68	69	70

4 tens < 5 tens < 6 tens

41 < 58 < 63

41, 58, 63

The numbers in the chart are in order from least to greatest.

The number 41 is in the first row.

The number 58 is in the next row.

The number 63 is in the last row.

41 is the **least** number.

63 is the **greatest** number.

Vocabulary

compare to look at objects, shapes, or numbers and see how they are alike or different

order 1, 3, 6, 7, 9 These numbers are in order from least to greatest.

The least number is 41. It comes first in a hundred chart. The greatest number is 63. It comes last in a hundred chart.

GO on

Order 73, 92, and 85 from **least** to **greatest**.

Step 1   Shade the numbers on a hundred chart.

71	72	73	74	75	76	77	78	79	80
81	82	83	84	85	86	87	88	89	90
91	92	93	94	95	96	97	98	99	100

Step 2   Start at 71. Move to the right. The least shaded number is 73.

Step 3   The next shaded number is 85.

Step 4   The greatest shaded number is 92.

Answer   73, 85, 92

Step-by-Step Practice

Order 66, 54, and 61 from **least** to **greatest**.

Step 1   Shade the numbers on a hundred chart.

51	52	53	54	55	56	57	58	59	60
61	62	63	64	65	66	67	68	69	70

Step 2   Start at _____. Move to the right.

The least shaded number is _____.

Step 3   The next shaded number is _____.

Step 4   The greatest shaded number is _____.

Answer   _____, _____, _____

Name _____

▶ Guided Practice

Order the numbers from **least** to **greatest**.

1 44, 39, 27

_____, _____, _____

21	22	23	24	25	26	27	28	29	30
31	32	33	34	35	36	37	38	39	40
41	42	43	44	45	46	47	48	49	50

2 35, 28, 39

_____, _____, _____

Problem-Solving Practice

3 Frosty Squares holds 22 ounces of cereal. Tasty Oats holds 32 ounces. Apple Crisps holds 25 ounces. Which box holds the greatest amount of cereal?

Understand Underline key words.

Plan Use a hundred chart.

Solve Shade the numbers.

21	22	23	24	25	26	27	28	29	30
31	32	33	34	35	36	37	38	39	40

The greatest number is _____.

_____ holds the greatest amount of cereal.

Check Use base-ten blocks to model the numbers.

GO on

▶ Practice on Your Own

Order the numbers from **least** to **greatest**.

51	52	53	54	55	56	57	58	59	60
61	62	63	64	65	66	67	68	69	70
71	72	73	74	75	76	77	78	79	80
81	82	83	84	85	86	87	88	89	90

4 57, 84, 63

_____, _____, _____

5 69, 80, 77

_____, _____, _____

6 59, 64, 53

_____, _____, _____

7 79, 86, 75

_____, _____, _____

8 66, 69, 62

_____, _____, _____

9 80, 87, 85

_____, _____, _____

---

10 **WRITING IN ▶MATH** Marta bought 51 animal stickers, 85 superhero stickers, and 56 flower stickers. Marta has the least amount of which kind of stickers? Explain.

_____

_____

_____

Vocabulary Check Complete.

11 The numbers 13, 15, and 17 are in _____ from least to greatest.

STOP

Name _____

Compare and Order Numbers to 500

Key Concept

You can use a place-value chart to **compare** and **order** numbers.

First, compare the hundreds.

Next, compare the tens.

Finally, compare the ones.

hundreds	tens	ones
1	5	4
1	6	2
1	8	8

The hundreds are all 1. Use the tens place.

5 tens < 6 tens < 8 tens

154 < 162 < 188

154, 162, 188

Vocabulary

compare   to look at objects, shapes, or numbers and see how they are alike or different

order   1, 3, 6, 7, 9
These numbers are in order from least to greatest.

154, 162, and 188 are alike because they have 1 hundred. They are different because they have different tens and ones.

GO on

Example

Order 341, 415, and 368 from **least** to **greatest**.

Step 1 Write the numbers in a place-value chart.

Step 2 Compare the hundreds.
4 hundreds > 3 hundreds
415 is the greatest number.

Step 3 Compare the tens in
341 and 368.
4 tens < 6 tens
341 < 368

Answer 341, 368, 415

hundreds	tens	ones
3	4	1
4	1	5
3	6	8

Step-by-Step Practice

Order 199, 172, and 267 from **least** to **greatest**.

Step 1 Write the numbers in a place-value chart.

Step 2 Compare the hundreds.

_____ hundreds $\bigcirc$ _____ hundred

_____ is the greatest number.

Step 3 Compare the tens in

_____ and _____.

_____ tens $\bigcirc$ _____ tens

_____ $\bigcirc$ _____

hundreds	tens	ones

Answer _____, _____, _____

Name _____

▶ Guided Practice

Order the numbers from **least** to **greatest**.

1 256, 116, 197

hundreds	tens	ones

_____ , _____ , _____

2 332, 424, 383

hundreds	tens	ones

_____ , _____ , _____

Problem-Solving Practice

3 The first grade collected 308 cans for a food drive. The second grade collected 284 cans. The third grade collected 261 cans. Which grade collected the most cans?

Understand Underline key words.

Plan Use a place-value chart.

Solve The greatest number

is _____.

hundreds	tens	ones

The _____ grade collected the most cans.

Check Use base-ten blocks to compare the numbers.

GO on

▶ Practice on Your Own

Order the numbers from **least** to **greatest**.

4 237, 134, 395

hundreds	tens	ones

———— , ———— , ————

5 419, 248, 311

———— , ———— , ————

6 371, 272, 345

———— , ———— , ————

7 108, 175, 134

———— , ———— , ————

8 432, 478, 453

———— , ———— , ————

9 ⬤ **WRITING IN ▶MATH** On Tuesday, 146 people visited a museum. On Wednesday, 292 people visited the museum. On Thursday, 217 people visited the museum. On which day did most people visit the museum? Explain.

_____

_____

_____

Vocabulary Check Complete.

10 I can _____ 422 and 341 to see which number is greater.

STOP

Name _____

Progress Check 3 (Lessons 3-5 and 3-6)

Order the numbers from **least** to **greatest**.

1 53, 69, 37

_____, _____, _____

31	32	33	34	35	36	37	38	39	40
41	42	43	44	45	46	47	48	49	50
51	52	53	54	55	56	57	58	59	60
61	62	63	64	65	66	67	68	69	70

2 46, 68, 64

_____, _____, _____

3 59, 53, 52

_____, _____, _____

4 297, 314, 252

hundreds	tens	ones

_____, _____, _____

5 416, 461, 428

hundreds	tens	ones

_____, _____, _____

6 Charlie read three books over the summer. The books have 65 pages, 38 pages, and 69 pages. Which book has the greatest number of pages? Explain.

_____

_____

Name _____

What is the Myst-purr-y Picture?

Solve each problem below.
Shade each answer orange in the picture.

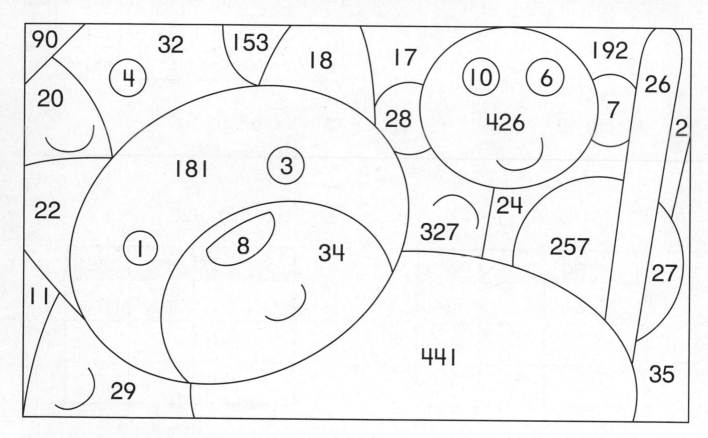

Circle the greatest number.

1 20 17 26   **2** 34 29 32   **3** 441 327 426

Circle the least number.

4 22 28 27   **5** 24 18 35   **6** 192 181 257

What does the picture show? a _____

128 one hundred twenty-eight

Name _____

Review

Vocabulary

Word Bank

equal to

greater than

less than

Use the Word Bank to complete.

1   <
    ⋮
    ↑ _____

2      >
    ⋮
    ↑ _____

3      =
    ⋮
    ↑ _____

▶ **Concepts**

Compare. Write >, <, or =.

4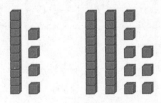

14 ⟨ < ⟩ 28

5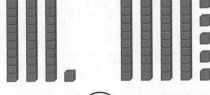

31 ◯ 45

Compare. Write >, <, or =.

6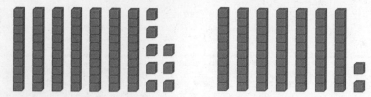

78 $\bigcirc$ 72

7 325 = __3__ hundreds __2__ tens __5__ ones

215 = _____ hundreds _____ ten _____ ones

325 $\bigcirc$ 215

8 549 = _____ hundreds _____ tens _____ ones

583 = _____ hundreds _____ tens _____ ones

549 $\bigcirc$ 583

Order the numbers from **least** to **greatest**.

9 74, 82, 78

__74__, _____, _____

71	72	73	74	75	76	77	78	79	80
81	82	83	84	85	86	87	88	89	90

10 89, 83, 86

_____, _____, _____

11 138, 189, 167

hundreds	tens	ones
1	3	8

_____, _____, _____

12 418, 434, 397

hundreds	tens	ones

_____, _____, _____

STOP

130 one hundred thirty

Chapter Test

Compare. Write >, <, or =.

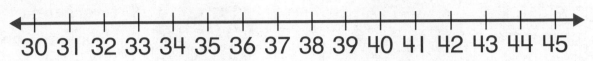

1   31 ◯ 41   **2**   39 ◯ 35   **3**   43 ◯ 43

4

   37 ◯ 42

5

   90 ◯ 81

6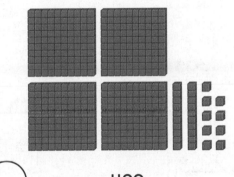

   412 ◯ 429

7   686 = _____ hundreds _____ tens _____ ones

   699 = _____ hundreds _____ tens _____ ones

   686 ◯ 699

8   Order 81, 96, and 86 from **least** to **greatest**.

| 81 | 82 | 83 | 84 | 85 | 86 | 87 | 88 | 89 | 90 |
| 91 | 92 | 93 | 94 | 95 | 96 | 97 | 98 | 99 | 100 |

_____, _____, _____

GO on

9 Who is Correct?

Carlos and Tia ordered 91, 98, and 88 from **least** to **greatest**.

From least to greatest, the numbers are 98, 91, 88.

Carlos

Tia

From least to greatest, the numbers are 88, 91, 98.

Circle the correct answer. Explain.

_____

_____

Order the numbers from **least** to **greatest**.

10 471, 483, 389

hundreds	tens	ones

_____, _____, _____

11 414, 432, 395

hundreds	tens	ones

_____, _____, _____

12 Sunee has 53 stamps in one album. The second album has 51 stamps. Which album has more stamps?

_____

_____

STOP

Test Practice

Listen as your teacher reads each problem.
Choose the correct answer.

1 Which number is least?

93   99   87   84
○   ○   ○   ○

2 Which symbol correctly compares the numbers?

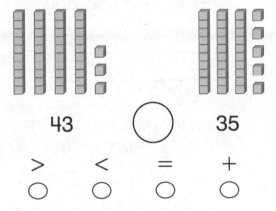

43   ◯   35

>   <   =   +
○   ○   ○   ○

3 Which comparison is correct?

○ 141 > 214
○ 214 > 141
○ 214 < 141
○ 214 = 141

4 The players on a soccer team compare the numbers on their jerseys. Which number is greatest?

30   28   31   29
○   ○   ○   ○

5 Maria has 25 stickers. Chad has 45 stickers. Which comparison is correct?

○ 25 > 45
○ 25 = 45
○ 25 < 45
○ 45 < 25

6 Marcus writes the following numbers. Which number does Marcus forget to write?

112, 113, ____, 115

111   114   116   131
○   ○   ○   ○

GO ON

7 Opa saved $245 in May, $159 in June, $312 in July, and $248 in August. Which month did Barbara save the greatest amount of money?

- ○ May
- ○ June
- ○ July
- ○ August

8 Which symbol correctly compares the numbers?

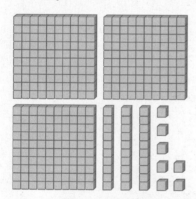

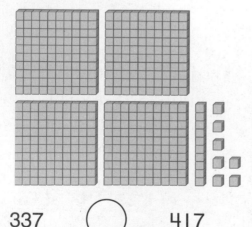

337 ◯ 417

>   <   =   +
○   ○   ○   ○

9 Which number is missing?

321, 322, ____, 324

320   323   333   325
○   ○   ○   ○

10 A baseball player got 159 hits, 187 hits, 205 hits, and 174 hits over four seasons. Which is the least number of hits?

- ○ 159 hits   ○ 187 hits
- ○ 205 hits   ○ 174 hits

11 Mrs. Castillo compares how many miles she drove on two road trips. Which comparison is correct?

- ○ 792 miles > 785 miles
- ○ 792 miles < 785 miles
- ○ 792 miles = 785 miles
- ○ 785 miles > 792 miles

12 Which symbol correctly compares the numbers?

852 ◯ 852

>   <   =   +
○   ○   ○   ○

STOP